SpringerBriefs in Materials

The SpringerBriefs Series in Materials presents highly relevant, concise monographs on a wide range of topics covering fundamental advances and new applications in the field. Areas of interest include topical information on innovative, structural and functional materials and composites as well as fundamental principles, physical properties, materials theory and design.

Indexed in Scopus (2022).

SpringerBriefs present succinct summaries of cutting-edge research and practical applications across a wide spectrum of fields. Featuring compact volumes of 50 to 125 pages, the series covers a range of content from professional to academic. Typical topics might include

- A timely report of state-of-the art analytical techniques
- A bridge between new research results, as published in journal articles, and a contextual literature review
- A snapshot of a hot or emerging topic
- An in-depth case study or clinical example
- A presentation of core concepts that students must understand in order to make independent contributions

Briefs are characterized by fast, global electronic dissemination, standard publishing contracts, standardized manuscript preparation and formatting guidelines, and expedited production schedules.

Milagrosa González Fernández de Castro
Juan José Moreno Labella
Miguel Panizo Laiz • Benito Del Río López

Alloyed Cast Iron

Springer

Milagrosa González Fernández de Castro
Escuela Técnica Superior de Ingenieros Industriales
Universidad Politécnica de Madrid (UPM)
Madrid, Spain

Juan José Moreno Labella
Escuela Técnica Superior de Ingenieros Industriales
Universidad Politécnica de Madrid (UPM)
Madrid, Spain

Miguel Panizo Laiz
Escuela Técnica Superior de Ingenieros Industriales
Universidad Politécnica de Madrid (UPM)
Madrid, Spain

Benito Del Río López
Escuela Técnica Superior de Ingenieros Industriales
Universidad Politécnica de Madrid (UPM)
Madrid, Spain

ISSN 2192-1091 ISSN 2192-1105 (electronic)
SpringerBriefs in Materials
ISBN 978-3-032-27302-4 ISBN 978-3-032-27303-1 (eBook)
https://doi.org/10.1007/978-3-032-27303-1

This Springer imprint is published by the registered company Springer Nature Switzerland AG
The registered company address is: Gewerbestrasse 11, 6330 Cham, Switzerland

Competing Interests The authors have no competing interests to declare that are relevant to the content of this manuscript.

Contents

Chapter 1
Theoretical Introduction to Alloyed Cast Irons

Abstract Cast irons are a broad family of multicomponent iron–carbon–silicon alloys that solidify through a eutectic reaction and display microstructures ranging from fully graphitic to fully carbide, depending on composition and cooling conditions (ASM Handbook Committee, ASM handbook, volume 1: properties and selection: irons, steels, and high-performance alloys, ASM International, 1990; Davis, ed., ASM specialty handbook: cast irons, ASM International, 1996; Angus, Cast iron: physical and engineering properties, 2nd ed., Butterworths, 1976; Laird et al., Abrasion-resistant cast iron handbook, American Foundry Society, 2000). Silicon additions strongly modify the Fe–C equilibrium diagram, shifting the eutectic and eutectoid transformations into temperature ranges and altering the solubility of carbon in austenite (ASM Handbook Committee, ASM handbook, volume 1: properties and selection: irons, steels, and high-performance alloys, ASM International, 1990; Angus, Cast iron: physical and engineering properties, 2nd ed., Butterworths, 1976). This behavior gives rise to two limiting solidification modes: the stable Fe–C–Si system, in which carbon precipitates as graphite, and the metastable Fe–Fe_3C system, where carbon remains combined as cementite, producing white irons (ASM Handbook Committee, ASM handbook, volume 1: properties and selection: irons, steels, and high-performance alloys, ASM International, 1990; Davis, ed., ASM specialty handbook: cast irons, ASM International, 1996; Angus, Cast iron: physical and engineering properties, 2nd ed., Butterworths, 1976; Laird et al., Abrasion-resistant cast iron handbook, American Foundry Society, 2000). Alloyed cast irons intentionally incorporate additional alloying elements beyond C, Si and Mn (with residual S and P) to tailor graphite morphology, carbide types and matrix constitution, thereby enabling a wide range of mechanical, tribological and high-temperature properties (Davis, ed., ASM specialty handbook: cast irons, ASM International, 1996; Angus, Cast iron: physical and engineering properties, 2nd ed., Butterworths, 1976; Laird et al., Abrasion-resistant cast iron handbook, American Foundry Society, 2000; Fashu et al., Metals 13(11):1831, 2023). They are commonly classified into low- or medium-alloy irons and high-alloy irons based on total alloying content, which is closely linked to their application domains and service environments (Davis, ed., ASM specialty handbook: cast irons, ASM International,

M. González Fernández de Castro et al., *Alloyed Cast Iron*, SpringerBriefs in Materials, https://doi.org/10.1007/978-3-032-27303-1_1

1996; Laird et al., Abrasion-resistant cast iron handbook, American Foundry Society, 2000; Fashu et al., Metals 13(11):1831, 2023).

Keywords Alloyed cast irons · Eutectic solidification · Fe–C–Si phase diagram · Graphite vs. cementite · Stable and metastable systems · High-alloy white irons · Abrasion-resistant irons

Cast irons are a broad family of multicomponent alloys whose base element is iron, characterized by having carbon and silicon contents higher than those of steels [1–4]. Metallurgically, their defining feature is that they solidify with the formation of a eutectic, a mixture of phases that solidifies at a nominally constant temperature [1, 2]. The presence of silicon in significant amounts profoundly modifies the iron–carbon equilibrium diagram, converting the eutectic and eutectoid transformation points into temperature ranges and affecting both the solubility of carbon and the position of the critical temperatures [1, 3].

The eutectic temperatures for austenite–graphite and austenite–cementite in Fe–C–Si cast irons can be approximated as a function of silicon content as [3]:

$$T_{\gamma-G} = 1159.014 + 2.6429(\%Si) + 0.4286(\%Si)^2$$

$$T_{\gamma-Fe3C} = 1150.171 - 9.4857(\%Si) - 2.8571(\%Si)^2$$

This gives rise to two possible solidification routes, depending on alloy composition, cooling rate and melt treatment [1–3]:

- **Stable System (Iron–Graphite)**: Carbon precipitates in its free form as graphite (Gr). The fracture of these cast irons is grayish in color, from which the name gray iron derives [1–3].
- **Metastable System (Iron–Iron Carbide)**: Carbon remains combined with iron in the form of iron carbide (Fe_3C or cementite). The fracture is white and crystalline, giving rise to white iron [1–4].

Alloyed cast irons are those to which other elements (in addition to C, Si, Mn, S, and P) are intentionally added to modify their microstructure and, consequently, their properties in a controlled manner [2–5]. From a compositional standpoint, they are usually classified into two large groups according to total alloying content: low- or medium-alloy irons, with a total alloying content below about 5 wt.% (a range that includes many pearlitic and martensitic white irons), and high-alloy irons, with higher alloying levels that promote the formation of complex carbides and special graphite morphologies for demanding abrasive and/or corrosive service [2, 4, 5] (Fig. 1.1).

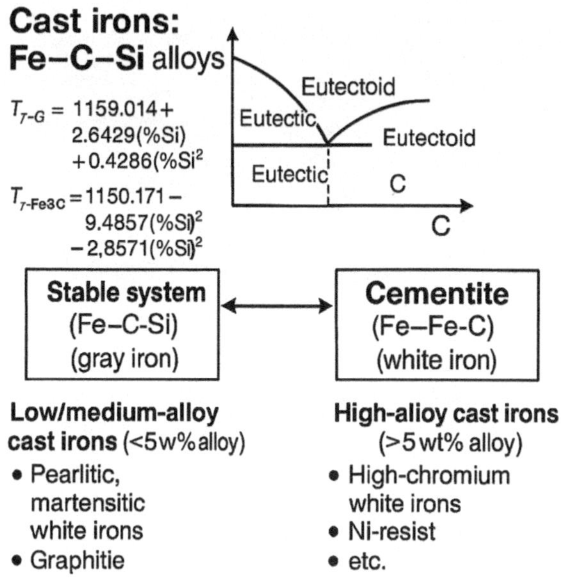

Fig. 1.1 Schematic/experimental representation of the main thermodynamic and microstructural features of alloyed cast irons, illustrating the influence of silicon content on the Fe–C–Si equilibrium diagram, the stable (austenite–graphite) and metastable (austenite–cementite) solidification paths, and their relationship with the resulting phase morphology and fracture characteristics

References

1. ASM Handbook Committee. (1990). *ASM handbook, volume 1: Properties and selection: Irons, steels, and high-performance alloys*. ASM International.
2. Davis, J. R. (Ed.). (1996). *ASM specialty handbook: Cast irons*. ASM International.
3. Angus, H. T. (1976). *Cast iron: Physical and engineering properties* (2nd ed.). Butterworths.
4. Laird, G., Gundlach, R., & Röhrig, K. (2000). *Abrasion-resistant cast iron handbook*. American Foundry Society.
5. Fashu, S., Xie, Y., Wang, J., & Xiao, W. (2023). Development and performance of high chromium white cast irons (HCWCIs) for wear/corrosive environments: A critical review. *Metals, 13*(11), 1831.

Chapter 2
Role of Graphite in Alloyed Cast Irons

Abstract Graphite morphology, distribution and size exert a dominant influence on the mechanical, physical and tribological behavior of alloyed cast irons (Davis, ed., ASM specialty handbook: cast irons, ASM International, Materials Park, OH, 1996; Stefanescu et al., in: ASM handbook, volume 1A: cast iron science and technology, ASM International, Materials Park, OH, 2017, pp. 169–211; Angus, Cast iron: physical and engineering properties, 2nd ed., Butterworths, London, UK, 1976; Laird et al., Abrasion-resistant cast iron handbook, American Foundry Society, Schaumburg, IL, 2000). Flake graphite in gray irons creates a connected network of internal notches that severely limits strength, ductility and toughness, but at the same time maximizes vibration damping and thermal conductivity (Davis, ed., ASM specialty handbook: cast irons, ASM International, Materials Park, OH, 1996; Stefanescu et al., in: ASM handbook, volume 1A: cast iron science and technology, ASM International, Materials Park, OH, 2017, pp. 169–211; Angus, Cast iron: physical and engineering properties, 2nd ed., Butterworths, London, UK, 1976). In contrast, spheroidal and aggregate graphite morphologies break continuous planes of weakness, allowing the metallic matrix to carry most of the load and thus providing "steel-like" combinations of strength, toughness and elongation (Davis, ed., ASM specialty handbook: cast irons, ASM International, Materials Park, OH, 1996; Stefanescu et al., in: ASM handbook, volume 1A: cast iron science and technology, ASM International, Materials Park, OH, 2017, pp. 169–211; Angus, Cast iron: physical and engineering properties, 2nd ed., Butterworths, London, UK, 1976; Tewary et al., Acta Mater 227:117725, 2022). Compacted (vermicular) graphite offers an intermediate balance between gray and ductile irons in terms of stiffness, damping and thermal properties (Stefanescu et al., in: ASM handbook, volume 1A: cast iron science and technology, ASM International, Materials Park, OH, 2017, pp. 169–211; König, Int J Cast Metals Res 23(2):73–80, 2010). Beyond morphology, graphite distribution (types A–E in gray irons) and graphite size strongly modulate performance, especially in applications governed by thermal fatigue, scuffing, abrasive wear and thermal-shock loading (Davis, ed., ASM specialty handbook: cast irons, ASM International, Materials Park, OH, 1996; Stefanescu et al., in: ASM handbook, volume 1A: cast iron science and technology, ASM International,

M. González Fernández de Castro et al., *Alloyed Cast Iron*, SpringerBriefs in Materials, https://doi.org/10.1007/978-3-032-27303-1_2

Materials Park, OH, 2017, pp. 169–211; Laird et al., Abrasion-resistant cast iron handbook, American Foundry Society, Schaumburg, IL, 2000; Roviglione et al., Minerals 11(2):109, 2021). Understanding the synergistic effect of morphology, distribution and size is therefore essential for the design and selection of alloyed cast irons tailored for demanding service conditions (Davis, ed., ASM specialty handbook: cast irons, ASM International, Materials Park, OH, 1996; Stefanescu et al., in: ASM handbook, volume 1A: cast iron science and technology, ASM International, Materials Park, OH, 2017, pp. 169–211; Angus, Cast iron: physical and engineering properties, 2nd ed., Butterworths, London, UK, 1976; Laird et al., Abrasion-resistant cast iron handbook, American Foundry Society, Schaumburg, IL, 2000; Tewary et al., Acta Mater 227:117725, 2022; König, Int J Cast Metals Res 23(2):73–80, 2010; Roviglione et al., Minerals 11(2):109, 2021).

Keywords Graphite morphology · Graphite distribution · Graphite size · Gray iron · Ductile iron · Compacted (vermicular) graphite iron · Mechanical properties · Vibration damping · Thermal conductivity

2.1 Graphite Morphology

The morphology and distribution of graphite govern the behavior of alloyed ferrous cast irons, often above the effect of the matrix itself. Graphite acts as a discontinuity: when it appears in lamellar form, the interconnected flakes constitute a network of internal notches that reduce strength, elongation, and toughness; in return, that same connectivity increases vibration damping and thermal conductivity. In contrast, in spheroidal or aggregate forms, the isolated nodules eliminate continuous plans of weakness and raise the yield strength and ductility, although they sacrifice part of the damping and conductivity. The distribution of lamellar graphite—types A to E—and certain anomalous forms such as Widmanstätten-type morphologies, mesh, "chunky," or intercellular flake graphite decisively modulate these responses. Process variables such as undercooling (D/E distributions), the hypereutectic condition (C distribution, "kish"), and nodule flotation in hypereutectic ductile iron also do so [1–4].

The microstructural basis is clear: graphite is an internal notch whose influence depends on its shape and degree of connectivity. A continuous network of interconnected flakes facilitates crack initiation and propagation and, although it penalizes tensile and impact response, it creates multiple graphite–matrix interfaces that provide high vibration damping and good thermal conduction performance; in addition, connectivity can favor phenomena such as graphitic corrosion by graphite acting as a cathode. Conversely, when graphite is present as isolated particles, preferred crack paths disappear, and the matrix can fully express its strength and toughness. The price is a reduction in damping and conductivity, precisely due to the loss of graphite continuity [1–4].

In lamellar graphite (gray iron), the microstructure is characterized by sharp, interconnected flakes whose acute geometry favors stress concentration. Mechanically, this translates into low strength and elongation and limited toughness, with a clearly asymmetric behavior between compression and tension, compressive strength being much higher than tensile strength. The massive presence of internal notches also explains a relatively low sensitivity to external notches: the microstructure is already dominated by internal notches. Physically, the connectivity of graphite confers very high vibration damping and thermal conductivity. Tribologically, lamellar morphology provides self-lubrication in dry sliding and generates cavities that act as oil reservoirs under lubrication. As a counterpart, the continuous graphite network can accelerate graphitic corrosion in which graphite acts as a cathode, as already mentioned [1–3, 5].

Spheroidal or nodular graphite consists of isolated nodules distributed throughout the matrix. Treatment with elements such as Mg or Ce is used to obtain this morphology. From the mechanical point of view, these morphologies allow much higher strengths, yield strengths, elongations, and toughnesses to be achieved, with a "steel-like" behavior; in ferritic ductile iron, elongation can exceed 20% [1–3, 6].

Aggregate graphite is described as irregular nodules or isolated rosettes. This morphology gives rise to a set of properties similar to nodular in terms of strength, ductility, and toughness. For its part, compacted or vermicular graphite occupies an intermediate position: morphologically it corresponds to short and relatively thick flakes with rounded ends; in properties, it offers a compromise between gray and ductile iron, improving the strength and ductility of gray iron while at the same time preserving damping and conductivity superior to those of nodular iron [2, 3, 7].

There are also various anomalous forms and degenerations. Widmanstätten-type and mesh morphologies are associated with the presence of traces such as lead or tellurium and entail catastrophic losses of strength. It is also important to note the appearance of "chunky" graphite and intercellular flake graphite as undesirable situations that degrade properties and consistency. These degenerations are linked to composition and processing conditions, and constitute risks that must be prevented due to their negative impact on performance [1–3, 5–7] (Fig. 2.1).

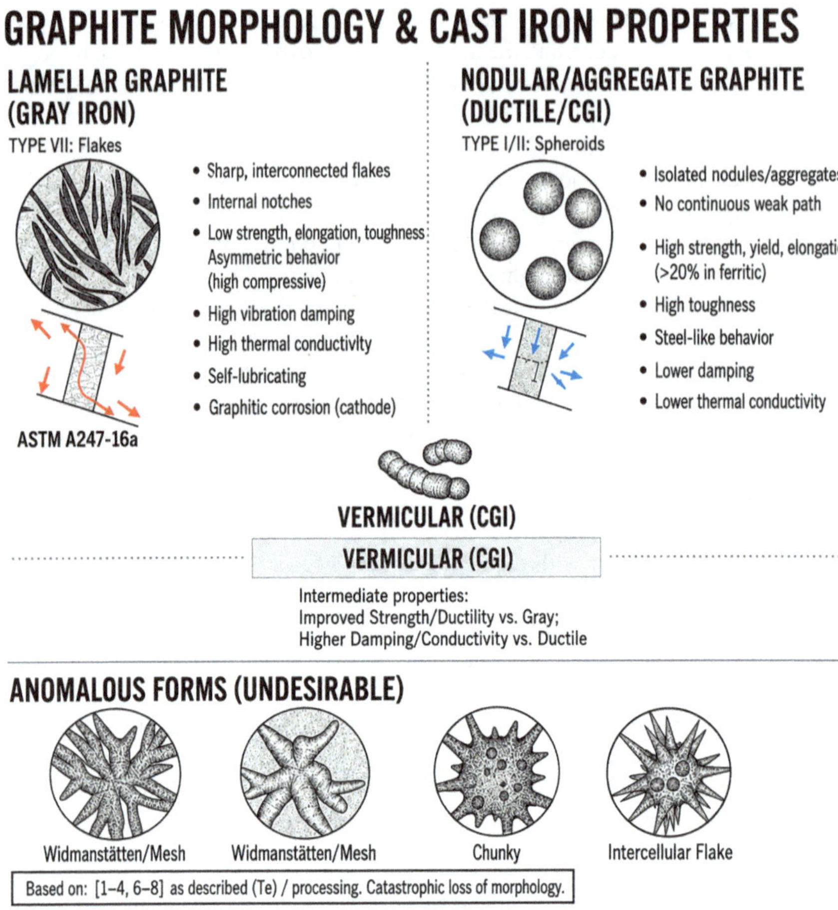

Fig. 2.1 Comparative schematic of the principal graphite morphologies in alloyed cast irons, including lamellar (gray iron), spheroidal (ductile iron), vermicular (compacted graphite iron), aggregate forms, and anomalous morphologies (Widmanstätten/mesh, chunky, intercellular flake). The diagram highlights differences in particle geometry, connectivity within the matrix, stress concentration behavior, and their qualitative influence on mechanical strength, toughness, damping capacity, and thermal conductivity

2.2 Graphite Distribution

As for distribution, the distribution of lamellar graphite is classified into five types (A–E) with different microstructural origins and differentiated consequences. Distribution A, uniform random, provides the best mechanical combination within gray cast irons. B, in rosettes, is associated with moderately rapid cooling. C corresponds to primary graphite ("kish") typical of hypereutectic compositions, characterized by thicker and longer flakes. Lastly, distributions D and E are typical of undercooling conditions and of thin sections or rapid cooling; in them, the fine, interdendritic graphite has a high specific surface that favors the formation of ferrite during the eutectoid transformation by reducing the necessary diffusion distances. The effect is a more ferritic matrix, although the baseline strength limitation persists due to the lamellar morphology itself [1–3, 5].

The interaction between graphite morphology and metallographic matrix defines the property framework. The operating principle is that the shape of the graphite sets the performance ceiling and the matrix fine-tunes within that framework. In gray iron, even if the matrix is hardened toward pearlite or martensite, the network of flakes keeps toughness and elongation at low levels; the comparative advantage lies in vibration damping and thermal conductivity [1–3, 5].

From the process point of view, solidification fixes the shape and distribution of the graphite, and heat treatments do not modify them substantially [1–3, 8].

Undercooling—linked to D/E—produces fine, interdendritic graphite whose large specific surface, again, promotes ferrite in the eutectoid. Hypereutectic conditions (C/kish) generate thick, long flakes; in hypereutectic nodular irons, nodule flotation may occur with zoning in properties. Traces or contaminants can induce undesirable morphologies such as the aforementioned Widmanstätten, mesh, or intercellular flake, with the consequent deterioration of performance [1, 2, 5, 8].

As for risks and defects, Widmanstätten and mesh morphologies should be avoided due to their catastrophic impact on strength, and "chunky" and intercellular flake configurations should be checked, as they compromise consistency and degrade performance. Undesired vermicular may reveal insufficient magnesium treatment. Distribution C (kish) leads to thick flakes with a negative impact on strength, and although D/E favors more ferritic matrices, the intrinsic limitation of the lamellar morphology persists. These observations fall within a practical guide: to maximize toughness, ductility, and impact resistance, spheroidal or aggregate morphologies are advisable, choosing the matrix—ferritic, pearlitic, ausferritic, or austenitic—according to the objective; to achieve maximum damping and high thermal conductivity, the lamellar morphology is preferable, assuming the mechanical limitations inherent to the flake network; for compromise solutions between strength/ductility and physical properties, compacted (vermicular) graphite offers an intermediate balance [2, 3, 5, 6] (Fig. 2.2).

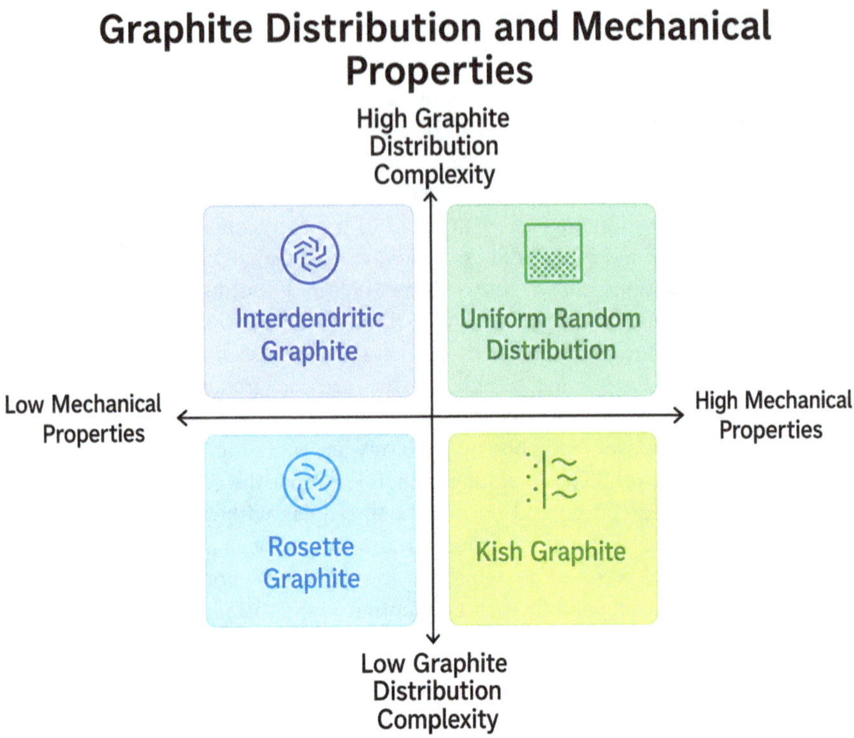

Fig. 2.2 Conceptual diagram relating graphite distribution types (interdendritic, uniform random, rosette, and kish graphite) to graphite distribution complexity and mechanical performance. The scheme qualitatively illustrates how different spatial arrangements of lamellar graphite influence strength and overall material behavior

2.3 Graphite Size

The size of the graphite is a crucial factor that, together with the composition and the microstructure of the matrix, determines the properties of graphitic cast irons. Precise control over the size of the graphite particles makes it possible to optimize the material for specific applications [1–5].

2.3.1 Factors That Control Graphite Size

The final size of the graphite particles in a casting is the result of the interaction between chemical composition and manufacturing process parameters [1–3, 5, 8].

2.3.1.1 Alloying Elements

Alloying elements have a direct effect on the refinement or coarsening of graphite:

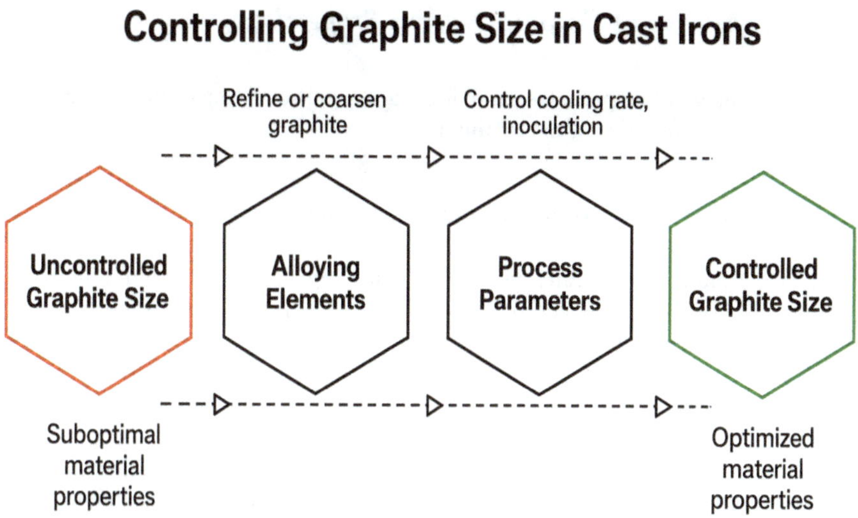

Fig. 2.3 Schematic representation of the main controlling factors of graphite size in cast irons, showing the interaction between alloying elements and process parameters (cooling rate and inoculation) and their effect on refining or coarsening graphite, ultimately determining optimized or suboptimal mechanical properties

- Elements that refine graphite (produce finer graphite):
  - Molybdenum (Mo): Has a strong refining effect on graphite.
  - Titanium (Ti): Is a powerful graphite refiner.
  - Vanadium (V): In amounts of 0.10–0.50%, refines the structure of the casting.
  - Nickel (Ni): Slightly refines the size of the graphite and the grain of the matrix.
  - Chromium (Cr): Slightly refines the graphite flakes.
  - Manganese (Mn): Also slightly refines the graphite.
- Elements that coarsen graphite (produce coarser graphite):
  - Carbon (C): Favors coarser graphite.
  - Silicon (Si): Tends to coarsen the graphite.
  - Aluminum (Al): Contributes to larger graphite.

2.3.1.2 Process Parameters

- Cooling Rate: It is a determining factor. Rapid cooling, such as that occurring in thin sections or in contact with a chill, produces finer graphite. Conversely, slow cooling, typical of thick sections, favors the formation of coarser graphite.
- Inoculation: The addition of certain elements just before pouring can influence grain size. For example, bismuth (Bi) can produce a fine-grained microstructure. Inoculation seeks to increase the number of graphitization nuclei, which promotes the formation of smaller eutectic grains and, therefore, finer graphite (Fig. 2.3).

2.3.2 *Influence of Graphite Size on Properties*

The size of the graphite particles has a direct and significant impact on the mechanical and service properties of the casting [1–5].

2.3.2.1 Mechanical Properties (Strength and Toughness)

- Finer graphite is generally associated with higher strength and toughness. Sources indicate that the addition of elements such as titanium (Ti) or vanadium (V) is carried out with the aim of decreasing the size of the graphite and the grain of the matrix to obtain high tensile strength (Rm).
- Nickel, which slightly refines the graphite, also helps increase the strength of the matrix.
- Conversely, coarser graphite tends to reduce the overall hardness of the material. Carbon, which coarsens the graphite, decreases transverse rupture strength.
- The overall refinement of the microstructure, including smaller graphite, tends to increase fracture toughness [1–5].

2.3.2.2 Wear Resistance

The influence of graphite size on wear is complex and depends on the type of wear:

- Scuffing: For this specific type of sliding wear, coarser graphite offers the highest resistance.
- Abrasive Wear: In applications where maximum abrasion resistance is sought, such as in white irons (Ni-Hard, high chromium), the presence of graphite, regardless of its size, is considered a serious defect that severely degrades performance. The goal in these alloys is for all the carbon to be in the form of hard carbides [2, 4, 8].

2.3.2.3 Vibration Damping

There is an inverse relationship between mechanical strength and damping capacity.

- Higher-strength cast irons (associated with finer graphite) tend to have lower damping capacity.
- Conversely, lower-strength materials (associated with coarser graphite) show a greater ability to absorb vibrations [1–3, 5, 7].

2.3.2.4 Thermal Conductivity

Thermal conductivity is related to the morphology (shape) of the graphite and the total amount present, rather than to its specific size. For example, the difference between lamellar (interconnected) graphite and spheroidal (isolated) graphite is

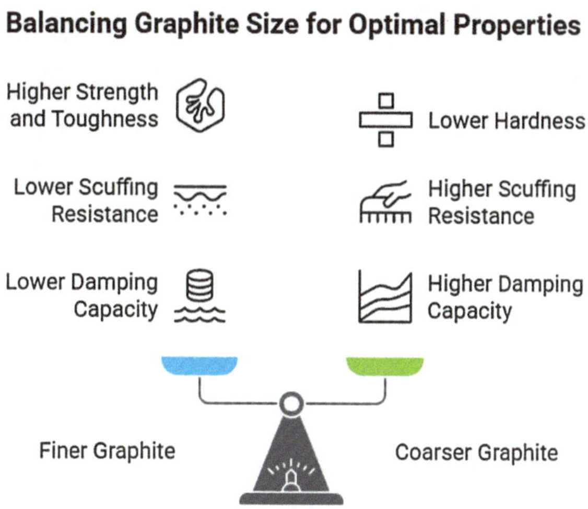

Fig. 2.4 Diagram illustrating the trade-off associated with graphite size, comparing finer and coarser graphite in terms of strength, toughness, hardness, scuffing resistance, and damping capacity. The balance concept emphasizes the inverse relationship between mechanical strength and vibration damping

highlighted, but there are no direct conclusions about how the size (fine vs. coarse) of the same morphology would affect this property [1–3, 5, 7].

In summary, control of graphite size is a fundamental metallurgical tool. Finer graphite is sought to maximize mechanical properties such as tensile strength and toughness. On the other hand, coarser graphite is beneficial for scuffing resistance and for greater vibration damping capacity. In the case of white irons designed to resist abrasion, any presence of graphite is detrimental [1–5] (Fig. 2.4).

2.4 Synergies Between Graphite Size, Morphology, and Distribution

In graphitic cast irons, the properties are not determined by a single characteristic of the graphite, but by the synergistic interaction of its morphology (shape), size, and distribution. These three factors are interconnected and are controlled by the chemical composition and the cooling rate. Their combined effect is what defines the performance of the material in service [1–5].

2.4.1 Interdependence of Graphite Characteristics

It is impossible to analyze size, shape, or distribution in isolation, since they influence each other during solidification:

- Cooling Rate: It is the most influential process factor. Rapid cooling promotes finer graphite with undercooled morphology (Type D), distributed in an interdendritic manner. Conversely, slow cooling favors coarser graphite, of lamellar morphology (Type A) and with a random distribution.

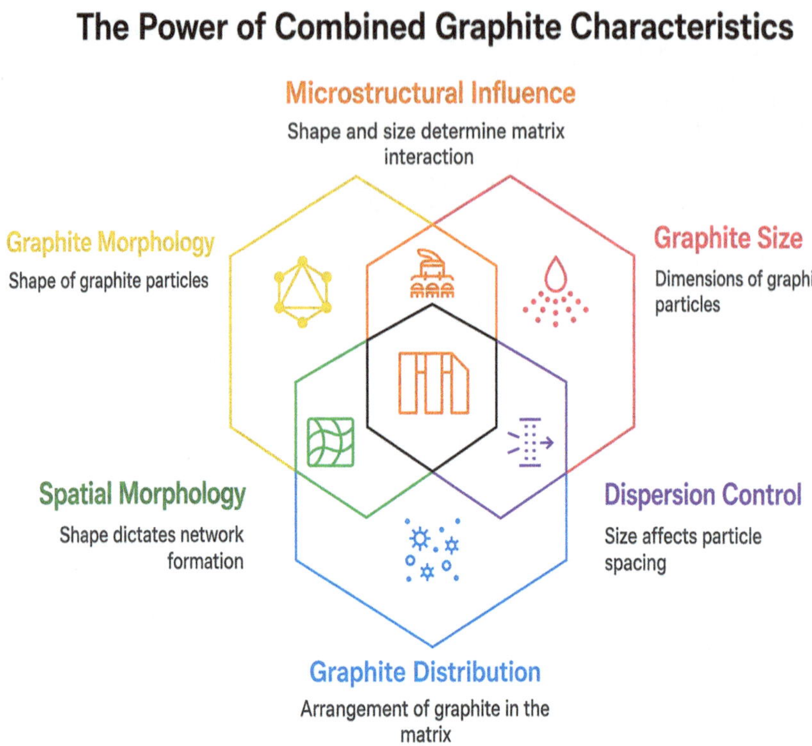

Fig. 2.5 Diagram illustrating the interdependence of graphite morphology, size, and distribution in cast irons. The scheme emphasizes how particle shape, dimensions, spatial arrangement, and dispersion collectively determine matrix interaction and ultimately control mechanical, physical, and tribological performance

- Chemical Composition: Inoculation with magnesium (Mg) or cerium (Ce) radically changes the morphology to spheroidal, which intrinsically creates a discrete and isolated distribution of the particles. The size of these spheroids can be controlled by the silicon content; an increase in Si can decrease the size of the spheroids by increasing the number of graphitization nuclei.
- Morphology and Distribution: The very shape of the graphite conditions its distribution. Lamellar graphite tends to form an interconnected network through the matrix, while spheroidal graphite, due to its globular nature, will always form a distribution of isolated particles [1–5] (Fig. 2.5).

2.4.2 Synergy and Its Influence on Key Properties

The interaction of these three graphite characteristics decisively defines the mechanical and physical properties of the casting.

2.4.2.1 Mechanical Properties (Strength, Toughness, and Ductility)

Here, morphology is the dominant factor, but its effect is amplified or mitigated by size and distribution.

- Lamellar Graphite (Alloyed Gray Irons):
  - Synergy: The sharp lamellar morphology, combined with an interconnected distribution (Type A), creates a network of stress concentrators that traverses the metallic matrix. A coarser flake size aggravates this effect, acting as a larger internal "crack."
  - Result: Low tensile strength and practically zero ductility. The material behaves in a brittle manner. The strength of gray iron is inversely proportional to the amount of free graphite.
  - Undercooled Graphite (Type D): Although it is also lamellar, its fine interdendritic distribution and its association with a ferritic matrix usually result in lower strength.
- Spheroidal Graphite (Alloyed Ductile Irons):
  - Synergy: The spherical morphology minimizes stress concentration. Its distribution as isolated particles allows the metallic matrix (ferritic, pearlitic, etc.) not to be interrupted and to carry the load continuously, expressing its inherent ductility and toughness. A smaller spheroid size and a higher nodule count (number of nodules per unit area) further improve these properties.
  - Result: High strength, toughness, and significant ductility, with properties that can be similar to those of steel.

2.4.2.2 Physical Properties (Thermal Conductivity and Vibration Damping)

In these properties, the distribution of the graphite plays a role as crucial as its morphology.

- Thermal Conductivity:
  - Synergy: Graphite is an excellent thermal conductor. In gray irons, the lamellar morphology and its interconnected distribution (Type A) create a continuous path for heat dissipation. This results in very high thermal conductivity.
  - In contrast, in ductile irons, the spheroidal morphology and its discrete and isolated distribution interrupt this thermal path. The metallic matrix, less conductive, dominates heat transport, resulting in significantly lower thermal conductivity.
- Vibration Damping (Damping Capacity):
  - Synergy: The lamellar morphology, being distributed as a network of discontinuities, allows the absorption of vibrational energy through internal friction at the graphite–matrix interfaces. A larger flake size and a greater volume of graphite increase this effect.

- Result: Gray irons have exceptionally high damping capacity. Conversely, the isolated distribution of spheroidal graphite drastically reduces this property, although it remains superior to that of steel.

2.4.2.3 Wear Resistance

The synergistic effect depends critically on the type of wear.

- Sliding Wear and Scuffing:
  - Synergy: It has been shown that a lamellar morphology of large size and random distribution (Type A) offers the maximum resistance. The coarse graphite is torn from the surface, creating cavities that act as lubricant reservoirs, while the exposed flakes act as a solid lubricant.
  - Result: Excellent scuffing resistance, superior to that of finer or spheroidal graphites.
- Abrasive Wear:
  - Synergy: In this case, any morphology, size, or distribution of graphite is detrimental, since it is a soft phase that is easily removed by abrasive particles.
  - Result: The presence of graphite is considered a serious defect in white irons designed to resist abrasion, since it severely degrades performance.

2.4.2.4 Thermal Shock Resistance

- Synergy: Gray irons with coarse lamellar graphite and randomly distributed (Type A) are superior. This is due to the combination of two factors:
  - High thermal conductivity (synergy of shape and distribution), which dissipates heat quickly and reduces thermal gradients.
  - Low modulus of elasticity (synergy of shape and distribution), which generates lower stresses for the same thermal deformation.
- Result: Good resistance to cracking by thermal shock, a property exploited in applications such as rolling mill rolls (Fig. 2.6).

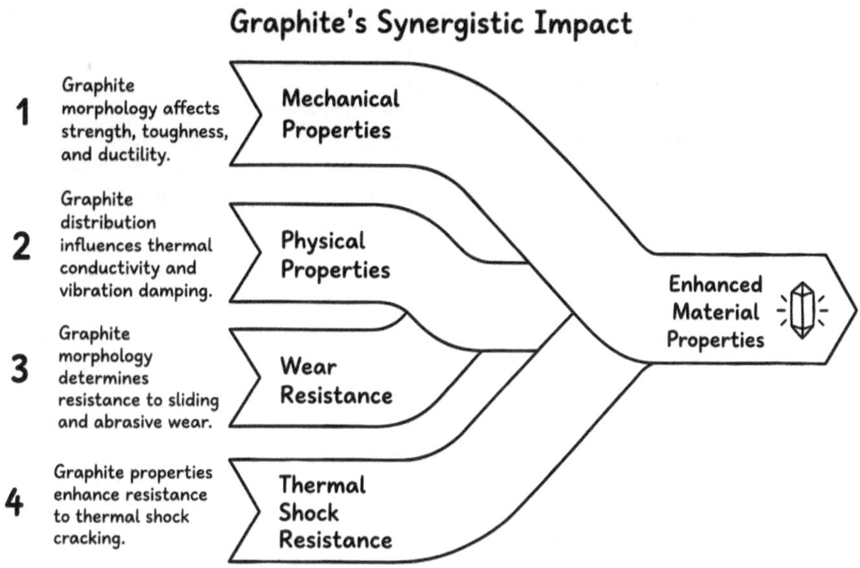

Fig. 2.6 Schematic summary of graphite's synergistic impact on cast iron properties, showing how morphology, distribution, and size influence mechanical properties, physical behavior, wear resistance, and thermal shock performance, leading to enhanced overall material characteristics

2.5 Synergistic Summary

In conclusion, to understand and design a graphitic cast iron, it is essential to consider the combined and synergistic effect of graphite size, morphology, and distribution, since it is this interaction that ultimately dictates the behavior and performance of the material [1–8] (Table 2.1).

Table 2.1 The table summarizes how the combination of graphite characteristics influences the key properties of cast irons

Property	Optimal morphology, distribution, and size	Justification of the synergy
Strength and toughness	Spheroidal, uniformly distributed, fine size	The spherical shape minimizes stress concentration, and its isolated distribution allows the matrix to carry the load without interruptions
Thermal conductivity	Lamellar, distributed in an interconnected network (Type A), coarse size	The lamellar morphology and network distribution create continuous paths for heat transfer through the graphite, which is highly conductive
Vibration damping	Lamellar, distributed in a network (Type A), coarse size	Graphite flakes distributed through the matrix absorb vibrational energy by internal friction at their interfaces. A larger size and volume of graphite maximize this effect
Scuffing resistance	Lamellar, distributed in a network (Type A), coarse size	Graphite acts as a solid lubricant. The cavities left by the coarse flakes serve as oil reservoirs
Thermal shock resistance	Lamellar, distributed in a network (Type A), coarse size	The combination of high thermal conductivity (dissipates heat) and low elastic modulus (reduces thermal stresses) is ideal

References

1. Davis, J. R. (Ed.). (1996). *ASM specialty handbook: Cast irons*. ASM International.
2. Stefanescu, D. M., Larrañaga, P., & de la Torre, U. (2017). Microstructures and characterization of gray irons. In *ASM handbook, volume 1A: Cast iron science and technology* (pp. 169–211). ASM International.
3. Angus, H. T. (1976). *Cast iron: Physical and engineering properties* (2nd ed.). Butterworths.
4. Laird, G., Gundlach, R., & Röhrig, K. (2000). *Abrasion-resistant cast iron handbook*. American Foundry Society.
5. Roviglione, A. N., et al. (2021). Graphite dendrites in cast iron and their fundamental role. *Minerals, 11*(2), 109. https://doi.org/10.3390/min11020109
6. König, M. (2010). Literature review of microstructure formation in compacted graphite iron. *International Journal of Cast Metals Research, 23*(2), 73–80. https://doi.org/10.1179/136404609X12535244328378
7. Sirtuli, L. J., Olsson, M., & Bushlya, V. (2024). Machining of compacted graphite iron: A review. *Journal of Materials Processing Technology, 319*, 118760. https://doi.org/10.1016/j.jmatprotec.2023.118760
8. Tewary, U., Stefanescu, D. M., et al. (2022). The origin of graphite morphology in cast iron. *Acta Materialia, 227*, 117725. https://doi.org/10.1016/j.actamat.2022.117725

Chapter 3
Purpose and Classification of Alloyed Cast Irons

Abstract The purpose of alloying cast irons is to deliberately manipulate the microstructure so as to obtain property combinations that cannot be achieved with unalloyed grades (ASM Handbook Committee, ASM handbook, volume 1: properties and selection: irons, steels, and high-performance alloys, ASM International, Materials Park, OH, 1990; Davis, ed., ASM specialty handbook: cast irons, ASM International, Materials Park, OH, 1996). Through controlled additions of alloying elements, cast irons can be designed to offer enhanced strength and hardness, improved wear and corrosion resistance, superior behavior at elevated temperatures, and tailored hardenability for thick and complex castings (ASM Handbook Committee, ASM handbook, volume 1: properties and selection: irons, steels, and high-performance alloys, ASM International, Materials Park, OH, 1990; Davis, ed., ASM specialty handbook: cast irons, ASM International, Materials Park, OH, 1996; Stefanescu et al., in: Stefanescu, ed., ASM handbook, volume 1A: cast iron science and technology, ASM International, Materials Park, OH, 2017, pp. 583–589; Laird et al., Abrasion-resistant cast iron handbook, American Foundry Society, Des Plaines, IL, 2000). From a metallurgical standpoint, the most useful classification of alloyed cast irons is based on the room-temperature microstructure, considering the presence or absence of free graphite and the nature of the metallic matrix (ASM Handbook Committee, ASM handbook, volume 1: properties and selection: irons, steels, and high-performance alloys, ASM International, Materials Park, OH, 1990; Davis, ed., ASM specialty handbook: cast irons, ASM International, Materials Park, OH, 1996; Stefanescu et al., in: Stefanescu, ed., ASM handbook, volume 1A: cast iron science and technology, ASM International, Materials Park, OH, 2017, pp. 583–589). Thus, alloyed cast irons with free graphite (gray, ductile and compacted graphite irons) are distinguished from alloyed white irons, in which carbon is combined in carbides and the fracture is bright and crystalline (ASM Handbook Committee, ASM handbook, volume 1: properties and selection: irons, steels, and high-performance alloys, ASM International, Materials Park, OH, 1990; Davis, ed., ASM specialty handbook: cast irons, ASM International, Materials Park, OH, 1996; Stefanescu et al., in: Stefanescu, ed., ASM handbook, volume 1A: cast iron science and technology, ASM International, Materials Park, OH, 2017, pp. 583–589; Laird

M. González Fernández de Castro et al., *Alloyed Cast Iron*, SpringerBriefs in Materials, https://doi.org/10.1007/978-3-032-27303-1_3

et al., Abrasion-resistant cast iron handbook, American Foundry Society, Des Plaines, IL, 2000). Within each group, subclasses are defined according to matrix type and primary application, such as pearlitic, bainitic, martensitic, ferritic and austenitic grades, or ledeburitic and ferritic high-chromium white irons (Davis, ed., ASM specialty handbook: cast irons, ASM International, Materials Park, OH, 1996; Laird et al., Abrasion-resistant cast iron handbook, American Foundry Society, Des Plaines, IL, 2000; Fashu and Trabadelo, Metals 13(11):1831, 2023). Finally, alloying elements are classified according to their graphitizing or carbide-forming tendency and their stabilizing effect on ferrite, pearlite or austenite, which underpins modern families of high-alloy cast irons used in demanding wear-, corrosion- and heat-resistant applications (ASM Handbook Committee, ASM handbook, volume 1: properties and selection: irons, steels, and high-performance alloys, ASM International, Materials Park, OH, 1990; Davis, ed., ASM specialty handbook: cast irons, ASM International, Materials Park, OH, 1996; Stefanescu et al., in: Stefanescu, ed., ASM handbook, volume 1A: cast iron science and technology, ASM International, Materials Park, OH, 2017, pp. 583–589; Laird et al., Abrasion-resistant cast iron handbook, American Foundry Society, Des Plaines, IL, 2000; Fashu and Trabadelo, Metals 13(11):1831, 2023).

Keywords Alloyed cast irons · Classification · Gray and white irons · Graphitizing and carbide-forming elements · Matrix stabilizers · High-chromium white irons · Ni-Resist and heat-resistant irons

The fundamental objective of alloying cast irons is to manipulate the microstructure to obtain properties that cannot be achieved with common cast irons. The main purposes are [1, 2]:

- Increase mechanical strength and hardness.
- Confer special properties such as high resistance to wear and abrasion, to corrosion (dry or electrochemical), and to heat (oxidation and growth).
- Modify hardenability to control the structure in parts of different thicknesses or to respond to heat treatments [1–3].

The most useful classification from a metallurgical point of view is based on the final microstructure at room temperature, taking into account the presence or absence of free graphite and the nature of the metallic matrix [1, 2, 4]:

3.1 Alloyed Cast Irons with Free Graphite

The fracture is dark. Graphite may be lamellar, spheroidal, compacted, etc., and the matrix may be [1, 2, 4]:

- Pearlitic: For high strength and good wear behavior (e.g., Ni-Tensil).
- Bainitic: For a combination of high strength and toughness (e.g., Acicular Iron).

- Martensitic: For high hardness obtained as-cast (e.g., Self-hardening Iron).
- Ferritic: For high resistance to oxidation or to acid corrosion (e.g., Silal, Duriron, Chugal).
- Austenitic: For resistance to corrosion and heat, or special physical properties (e.g., Ni-Resist, Minvar, Nomag) [1–3].

3.2 Alloyed Cast Irons Without Free Graphite (White)

The fracture is white and bright. They contain no graphite, and their microstructure consists of hard carbides with a metallic or ferritic matrix [1–3, 5].

- Ledeburitic: For exceptional abrasion resistance (e.g., Ni-Hard, High-Chromium Irons).
- Ferritic: For extreme resistance to oxidation at very high temperatures (e.g., irons with >28% Cr or ≈7% Al) [1–3, 5] (Fig. 3.1).

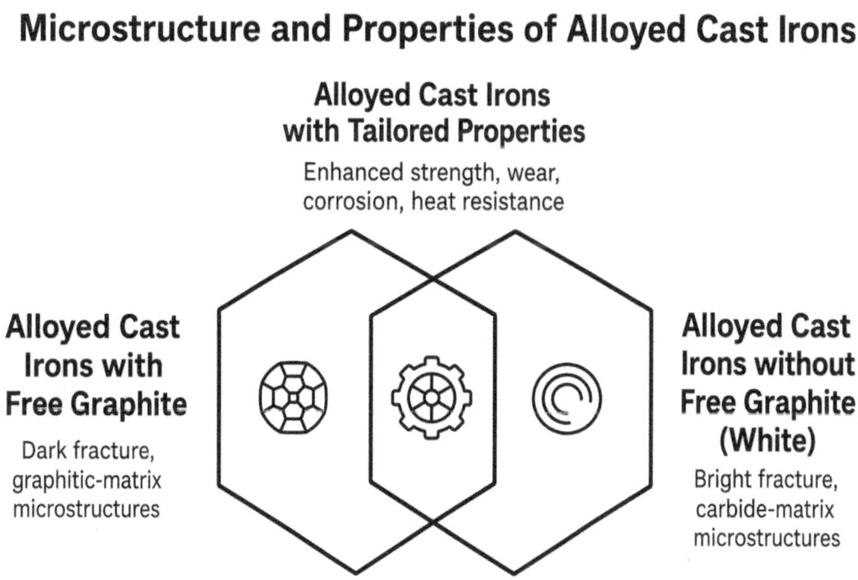

Fig. 3.1 Schematic diagram of the metallurgical classification of alloyed cast irons based on the presence or absence of free graphite and the resulting fracture appearance. The figure contrasts alloyed cast irons with free graphite (dark fracture, graphitic microstructures) and alloyed white irons without graphite (bright fracture, carbide-dominated microstructures), emphasizing the relationship between microstructure and tailored properties

3.3 The Role of Alloying Elements

Each alloying element has a complex and often interdependent effect on the microstructure. Its influence can be understood through two main mechanisms: its effect on the graphitizing potential and its effect on the stability of the matrix phases [1, 2].

- *Graphitizing Potential*: Refers to the tendency of the casting to solidify via the stable system (with graphite) instead of the metastable one (with carbides). Elements are classified according to how they affect the temperature of the stable and metastable eutectics:
  - Graphitizing Elements (Graphite Stabilizers): Such as silicon, aluminum, nickel, and copper. They increase the graphitizing potential, promoting the formation of graphite and preventing white solidification (chill).
  - Carbide-Forming Elements (Carbide Stabilizers): Such as chromium, molybdenum, vanadium, and tungsten. They decrease the graphitizing potential, promoting the formation of carbides and hindering the appearance of graphite [1, 2, 4].
- *Effect on the Matrix*: The alloying elements also determine the microstructure of the metallic matrix that forms during cooling in the solid state, mainly through their influence on the eutectoid transformation of austenite [1–4].
  - Ferrite Stabilizers (Alpha-Formers): Silicon, aluminum, and chromium (at high concentrations) raise the critical temperature (A1), promoting a ferritic matrix stable at high temperatures. This is fundamental in heat-resistant irons such as Silal or high-chromium irons (>28% Cr) [1–4].
  - Pearlite Stabilizers: Manganese (in excess), chromium, molybdenum, copper, nickel, and tin delay the decomposition of austenite into ferrite, favoring the formation of a harder, stronger pearlitic matrix, or refining its structure [1–4].
  - Austenite Stabilizers (Gamma-Formers): Nickel, manganese, and copper, in sufficiently high concentrations, depress both the pearlitic and martensitic transformation temperatures below room temperature. This allows an austenitic matrix to be obtained, as in Ni-Resist irons.
  - Increased Hardenability: Molybdenum, nickel, copper, manganese, and chromium shift the transformation (TTT) curves to the right, slowing the transformation of austenite. This makes it possible to obtain bainitic or martensitic matrices with slower cooling rates, which is key for Acicular Irons and Ni-Hard [1–5].

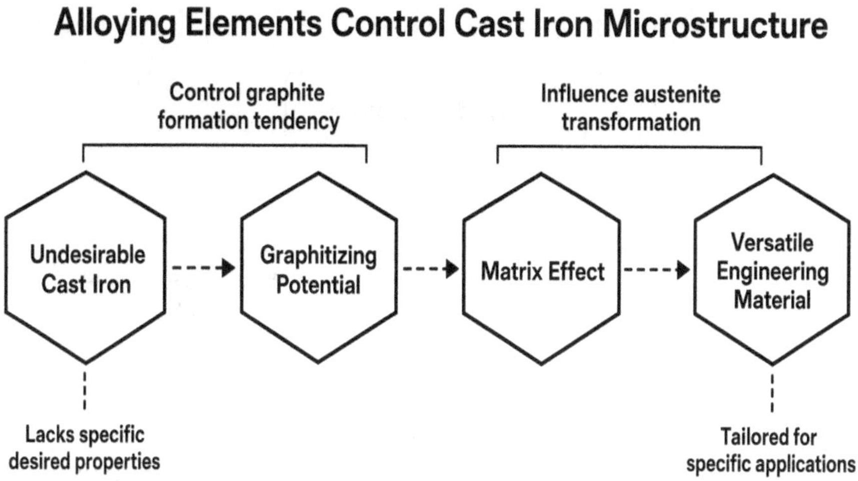

Fig. 3.2 Conceptual diagram illustrating how alloying elements control cast iron microstructure through two primary mechanisms: modification of graphitizing potential (stable vs. metastable solidification) and influence on austenite transformation and matrix development. The scheme highlights the transition from undesirable cast iron to a versatile engineering material tailored for specific applications

In summary, the addition of alloying elements transforms cast iron into a highly versatile engineering material. By precisely controlling the chemical composition, it is possible to design alloys with specific microstructures—whether by the shape of the graphite, the nature of the matrix, or the presence and type of carbides—to meet extreme demands of mechanical performance, wear, corrosion, or heat resistance. Knowledge of these theoretical principles makes it possible to select or develop the most suitable alloyed cast iron for each industrial application, from high-strength engine components to parts for chemical plants or mining equipment subjected to maximum abrasion [1–5] (Figs. 3.2 and 3.3).

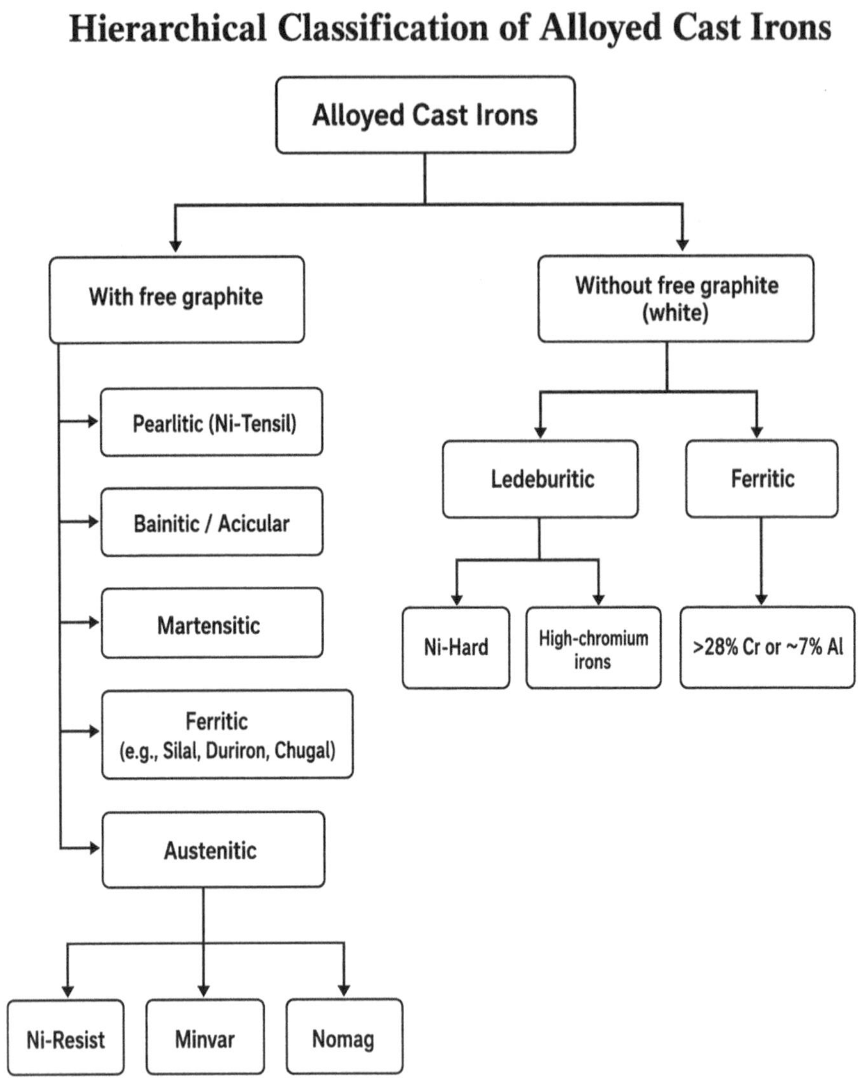

Fig. 3.3 Hierarchical classification chart of alloyed cast irons showing families with free graphite (pearlitic, bainitic/acicular, martensitic, ferritic, austenitic grades such as Ni-Resist) and without free graphite (Ni-Hard, high-chromium, high-aluminum irons), illustrating their subdivision according to matrix type and primary application

References

1. ASM Handbook Committee. (1990). *ASM handbook, volume 1: Properties and selection: Irons, steels, and high-performance alloys*. ASM International.
2. Davis, J. R. (Ed.). (1996). *ASM specialty handbook: Cast irons*. ASM International.
3. Laird, G., Gundlach, R., & Röhrig, K. (2000). *Abrasion-resistant cast iron handbook*. American Foundry Society.
4. Stefanescu, D. M., Larrañaga, P., & de la Torre, U. (2017). Microstructures and characterization of gray irons. In D. M. Stefanescu (Ed.), *ASM handbook, volume 1A: Cast iron science and technology* (pp. 583–589). ASM International.
5. Fashu, S., & Trabadelo, V. (2023). Development and performance of high chromium white cast irons (HCWCIs) for wear–corrosive environments: A critical review. *Metals, 13*(11), 1831. https://doi.org/10.3390/met13111831

Chapter 4
The Role of Alloying Elements in Ferrous Cast Irons

Abstract Alloying elements are added to cast irons in order to deliberately manipulate the microstructure and thereby obtain combinations of properties that cannot be achieved with unalloyed grades (ASM Handbook Committee, ASM handbook, volume 1: properties and selection: irons, steels, and high-performance alloys, ASM International, Materials Park, OH, 1990; Davis, ed., ASM specialty handbook: cast irons, ASM International, Materials Park, OH, 1996). Through controlled additions of graphitizing and carbide-forming elements, together with matrix stabilizers, cast irons can be designed to provide enhanced strength and hardness, improved wear and corrosion resistance, superior performance at elevated temperatures, and tailored hardenability for thick and complex castings (ASM Handbook Committee, ASM handbook, volume 1: properties and selection: irons, steels, and high-performance alloys, ASM International, Materials Park, OH, 1990; Davis, ed., ASM specialty handbook: cast irons, ASM International, Materials Park, OH, 1996; Stefanescu, ed., ASM handbook, volume 1A: cast iron science and technology, ASM International, Materials Park, OH, 2017; Laird et al., Abrasion-resistant cast iron handbook, American Foundry Society, Des Plaines, IL, 2000). From a metallurgical standpoint, a useful way to classify alloyed cast irons is to distinguish between irons with free graphite and white irons in which carbon is combined in carbides, and then to further subdivide them according to matrix type and primary application (pearlitic, bainitic, martensitic, ferritic and austenitic grades, as well as ledeburitic and high-chromium white irons) (ASM Handbook Committee, ASM handbook, volume 1: properties and selection: irons, steels, and high-performance alloys, ASM International, Materials Park, OH, 1990; Davis, ed., ASM specialty handbook: cast irons, ASM International, Materials Park, OH, 1996; Stefanescu, ed., ASM handbook, volume 1A: cast iron science and technology, ASM International, Materials Park, OH, 2017; Laird et al., Abrasion-resistant cast iron handbook, American Foundry Society, Des Plaines, IL, 2000). The present chapter reviews the role of the most important alloying elements in cast irons, including their influence on graphite formation versus carbide precipitation, their stabilizing effect on ferrite, pearlite or austenite, and a number of technologically relevant synergies that underpin modern families of abrasion-, corrosion- and heat-resistant cast

M. González Fernández de Castro et al., *Alloyed Cast Iron*, SpringerBriefs in Materials, https://doi.org/10.1007/978-3-032-27303-1_4

irons (ASM Handbook Committee, ASM handbook, volume 1: properties and selection: irons, steels, and high-performance alloys, ASM International, Materials Park, OH, 1990; Davis, ed., ASM specialty handbook: cast irons, ASM International, Materials Park, OH, 1996; Stefanescu, ed., ASM handbook, volume 1A: cast iron science and technology, ASM International, Materials Park, OH, 2017; Laird et al., Abrasion-resistant cast iron handbook, American Foundry Society, Des Plaines, IL, 2000; Fashu et al., Metals 13(11):1831, 2023).

Keywords Alloying elements · Cast irons · Graphitizers · Carbide-forming elements · Matrix stabilizers · High-chromium white irons · Ni-Resist · Wear- and heat-resistant irons

Alloying elements are added to cast irons to deliberately modify the microstructure and, consequently, the properties of the final material. Their influence is complex, since each element can simultaneously affect the tendency to graphitize, the structure of the matrix (ferrite, pearlite, austenite, etc.), the morphology of the graphite, and the mechanical and physical properties. Often, the greatest benefits are obtained by using combinations of alloying elements to leverage their synergistic effects.

What follows details the role of each alloying element [1, 2].

4.1 Main Elements (Besides Iron)

4.1.1 Carbon (C)

Carbon is the most important element after iron, and its content fundamentally defines a cast iron. In commercial cast irons, its percentage usually varies between 2.5% and 4.0%. Its main influence lies in the fact that, in amounts above the maximum solubility in austenite, it precipitates as a carbon-rich phase, which can be graphite (stable system) or iron carbide/cementite (metastable system).

- Effect on the microstructure: A higher carbon content increases the amount of graphite or iron carbide, which in turn influences almost all properties. A high carbon content in austenite lowers the start temperature of the martensitic transformation (Ms). Carbon dissolved in martensite is chiefly responsible for its hardness.
- Effect on mechanical properties: A higher total carbon content is generally associated with lower tensile strength and a lower elastic modulus, due to the greater fraction of graphite. However, in white irons, high carbon increases hardness and abrasion resistance by increasing the fraction of carbides, albeit at the expense of toughness [1, 2].
- Effect on physical and casting properties: Increases fluidity and improves castability. Increasing carbon content can decrease thermal conductivity due to the

increase of low-conductivity microconstituents such as pearlite and cementite [1, 2].

4.1.2 Silicon (Si)

Silicon is the second most important element and acts as a powerful graphitizer, favoring the formation of graphite over iron carbide.

- Effect on the microstructure: It is an alpha-forming element, i.e., it stabilizes ferrite. It raises the eutectoid transformation temperature (critical α–γ), which is fundamental in heat-resistant irons such as Silal. In gray irons, a high silicon content promotes a ferritic matrix and coarser graphite. In white irons, it must be kept low to avoid graphitization, but in high-chromium irons, it can promote the formation of martensite by raising the Ms temperature [1–3].
- Effect on mechanical properties: By promoting ferrite, it generally softens the iron and reduces its strength, unless it is dissolved in ferrite at high concentrations, in which case it hardens it significantly (silico-ferrite). It increases brittleness, especially in high-silicon ferritic cast irons such as Duriron and Silal, and raises the impact transition temperature [1–3].
- Effect on physical and casting properties: Increases casting fluidity. Reduces density. Notably increases electrical resistivity. Thermal conductivity is affected in a complex way: silicon dissolved in the matrix reduces it, but by promoting graphite (which is a good conductor), the net effect may vary [1–3].
- Resistance to corrosion and heat: It is a key element for oxidation resistance at high temperatures, by forming a protective surface layer of silicon oxide (SiO_2) or iron silicate. It is the fundamental alloying element in Silal (4–6% Si) and Duriron (14.5% Si) cast irons [1–4].

4.1.3 Manganese (Mn)

Manganese is a carbide stabilizer and a promoter of pearlite. Its main function is to neutralize sulfur by forming manganese sulfide (MnS), which is less harmful than iron sulfide. Sulfur is considered neutralized when $\%Mn \geq 1.7 \times \%S + 0.3\%$.

- Effect on the microstructure: Excess manganese (not combined with sulfur) refines pearlite, increasing its amount and hardness. It is an austenite stabilizer (gamma-former), thus depressing the critical temperature. At high concentrations, it increases the amount of retained austenite in martensitic cast irons, which can be detrimental to hardness and abrasion resistance [1–3].
- Effect on mechanical properties: Increases hardness and strength by stabilizing pearlite. Also increases hardenability.

- Effect on physical properties: Decreases the viscosity of the molten metal. Decreases thermal conductivity. Reduces magnetic permeability [1–3].
- Applications: Used to partially replace nickel in low-cost, non-magnetic austenitic cast irons such as Nomag (~5% Mn). In Ni-Hard, it is controlled so as not to stabilize too much austenite [1–3].

4.1.4 *Phosphorus (P)*

Phosphorus forms a low-melting eutectic (approx. 930 °C) known as steadite, which solidifies in interdendritic regions.

- Effect on the microstructure: Acts as a graphitizer, reducing the carbon content of the eutectic. Steadite is a very hard constituent (400–600 HB) composed of iron phosphide (Fe_3P) and ferrite. From 0.4–0.7% P, it forms an almost continuous intercellular network [1–4].
- Effect on mechanical properties: Notably increases hardness (30–40 HB per 1% P) and resistance to sliding wear and scuffing. However, it increases brittleness and drastically reduces impact strength [1–4].
- Effect on physical and casting properties: Significantly improves castability (fluidity) of the molten metal, making it ideal for thin and complex parts such as radiators. Increases electrical resistivity above 0.2% [1–4].
- Applications: Widely used in gray irons for automotive (liners, rings) and decorative parts where fluidity and wear resistance are key. Kept low (<0.08%) in high-toughness cast irons or for high-temperature service such as Ni-Resist [1–4].

4.1.5 *Sulfur (S)*

Sulfur is generally considered a harmful impurity. It is a strong carbide stabilizer and promotes white solidification (chill) [1, 2].

- Effect on the microstructure: If not neutralized by manganese, it forms iron sulfide (FeS) at grain boundaries, causing hot shortness. A very low sulfur content (<0.03%) can promote the formation of undesirable (interdendritic) graphite instead of Type A graphite [1, 2].
- Effect on properties: Reduces strength and ductility. Degrades abrasion resistance [1, 2].
- Effect on physical properties: Increases the viscosity of the molten metal. Markedly reduces surface tension [1, 2].

4.1.6 Nickel (Ni)

Nickel is a very versatile element with a dual effect: it is a moderate graphitizer and a powerful austenite stabilizer (gamma-former) [1–3].

- Effect on the microstructure:
  - As a graphitizer (approx. 1/3 the effectiveness of silicon), it prevents white solidification (chill) in thin sections and corners, improving machinability.
  - As an austenite stabilizer, it refines pearlite, increasing the hardness and strength of the matrix in low-alloy cast irons (e.g., Ni-Tensil). At higher concentrations (3.3–6.5%), it suppresses the pearlitic transformation and allows a martensitic matrix to be obtained in the as-cast state (e.g., Ni-Hard). With still higher contents (>13%), it stabilizes austenite down to room temperature, creating austenitic cast irons (e.g., Ni-Resist, Nomag) [1–3].
- Effect on mechanical properties: Increases strength, hardness, and toughness. Austenitic cast irons with nickel have excellent toughness at cryogenic temperatures, since they do not show a ductile–brittle transition [1–3].
- Effect on physical properties: Depresses the critical temperature. Decreases thermal conductivity and increases electrical resistivity. Notably affects the coefficient of thermal expansion, allowing very low-expansion alloys to be created (Minvar, with 34–36% Ni). Decreases magnetic permeability, and at high concentrations makes the cast iron non-magnetic [1–3].
- Resistance to corrosion and heat: Increases corrosion resistance in alkaline and reducing media. It is the key alloying element in the Ni-Resist family. It should not be used above 575 °C in sulfur-bearing atmospheres [1–3].

4.1.7 Chromium (Cr)

Chromium is a powerful carbide stabilizer and strongly opposes graphitization [1–5].

- Effect on the microstructure [1–5]:
  - Forms chromium carbides that are very hard and stable at high temperature. With $Cr < 10\%$, it forms M_3C carbides (alloyed cementite). Above 10–12%, it forms M_7C_3 carbides, which are harder (1300–1800 HV) and of discontinuous morphology, improving toughness. With $Cr > 30\%$, it forms $M_7C_3/M_{23}C_6$ carbides depending on composition and heat treatment.
  - Refines pearlite and increases matrix hardness. At high concentrations (>28% Cr), it stabilizes a fully ferritic matrix [1–5].

- Effect on mechanical properties: Drastically increases hardness, wear resistance, and hot strength. Increases hardenability [1–5].
- Effect on physical properties: Raises the critical temperature. Its effect on thermal conductivity is reported inconsistently [1–5].
- Resistance to corrosion and heat: The key element for oxidation resistance at high temperatures and for corrosion, especially in oxidizing media, by forming a passive chromium oxide film [1–5].
- Applications: Used in small amounts to stabilize pearlite and improve wear in gray irons. An essential component in Ni-Hard cast irons (to ensure white solidification) and the main element in high-chromium cast irons (11–35% Cr) [1–5].

4.1.8 Molybdenum (Mo)

Molybdenum is a moderate carbide stabilizer and a powerful agent for increasing hardenability.

- Effect on the microstructure: Very effectively suppresses the formation of pearlite, even in thick sections, without excessively stabilizing retained austenite. Strongly refines graphite and pearlite.
- Effect on mechanical properties: Significantly increases strength, hardness, and toughness, especially in acicular cast irons, where it is a key alloying element. It is the element that most increases creep resistance and rupture strength at elevated temperatures.
- Effect on physical properties: Its addition to Durichlor cast iron specifically improves resistance to hydrochloric acid.
- Applications: Used in high-strength cast irons (Ni-Tensil, Acicular), in Ni-Hard and high-chromium cast irons to improve hardenability in thick sections, and in Si-Mo and austenitic ductile irons to improve high-temperature performance [1, 2, 4, 5].

4.1.9 Copper (Cu)

Copper acts as a mild graphitizer and a promoter of pearlite.

- Effect on the microstructure: Refines pearlite and increases the strength and hardness of the matrix. At high concentrations, it is an austenite stabilizer.
- Effect on mechanical properties: Increases tensile strength and wear resistance. Provides anti-friction qualities.

- Effect on physical and casting properties: Increases castability. Improves resistance to atmospheric corrosion and to weak acids. Reduces magnetic permeability and increases hysteresis loss.
- Limitations: Its presence hinders graphite nodularization, so its use in ductile irons is limited or requires very strict control.
- Applications: Often used as a more economical substitute for nickel in high-strength pearlitic cast irons. It is a component of some Ni-Resist alloys [1, 2, 4, 5].

4.1.10 Aluminum (Al)

Aluminum is a complex element that acts as a graphitizer and ferrite stabilizer (alpha-former), but it can also stabilize cementite in certain concentration ranges.

- Effect on the microstructure: Its graphitizing action is strong up to 2%, but decreases between 2% and 4%. Between 7% and 18%, it stabilizes cementite, promoting a white-iron structure. At high concentrations (20–24%), it again becomes a powerful graphitizer, giving rise to Chugal cast iron.
- Effect on mechanical properties: In graphitic cast irons, it increases tensile strength, especially if the graphite is spheroidal. In white irons, it produces hard but brittle alloys.
- Effect on physical properties: Markedly reduces density and thermal conductivity. Raises the pearlitic transformation temperature. With more than 20% Al, the alloy becomes paramagnetic.
- Resistance to corrosion and heat: An exceptional element for oxidation resistance at high temperatures (up to 950–1100 °C), by forming a protective layer of aluminum oxide (Al_2O_3).
- Production challenges: Very difficult to cast due to its high reactivity with oxygen, which causes slag formation and inclusions. It can form aluminum carbide (Al_4C_3), which reacts with moisture and causes disintegration of the part [1, 2, 4, 5] (Fig. 4.1).

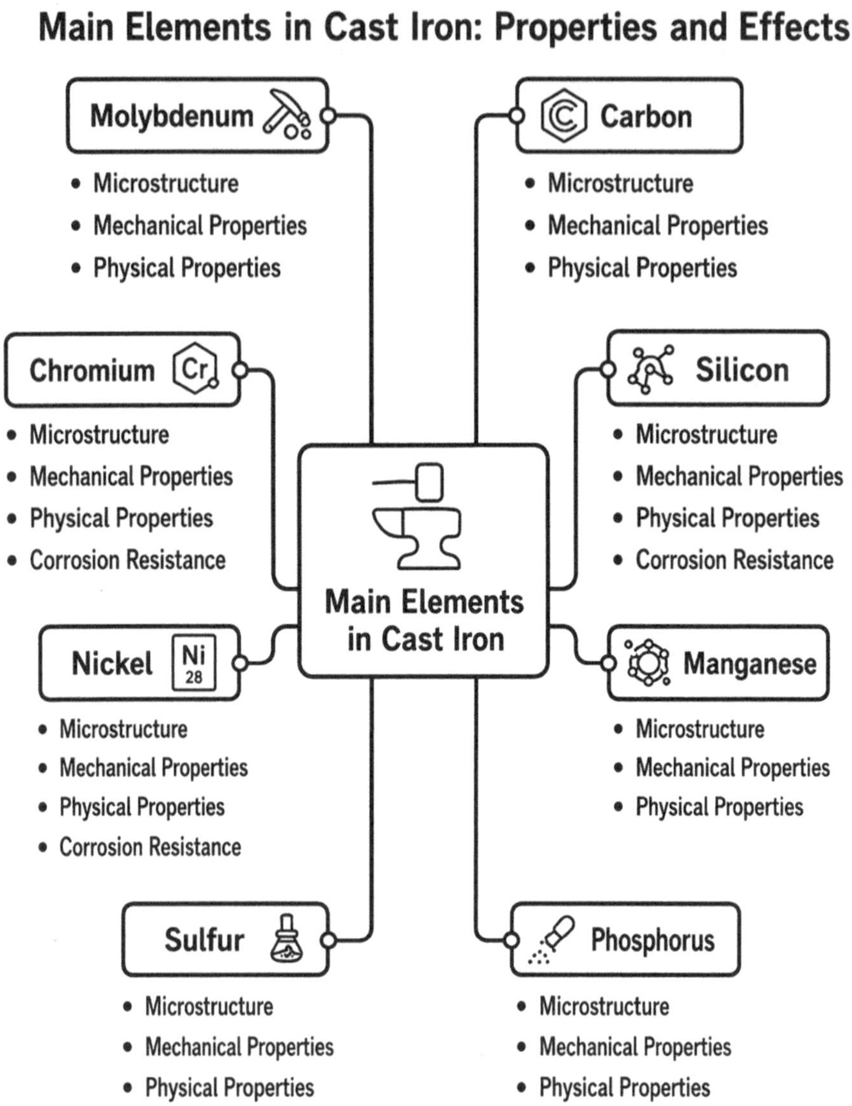

Fig. 4.1 Conceptual diagram summarizing the role of the main alloying elements in cast irons (C, Si, Mn, P, S, Ni, Cr, Mo), illustrating their influence on microstructure, mechanical properties, physical properties, and corrosion resistance, and emphasizing their interconnected metallurgical effects

4.2 Minor and Trace Alloying Elements

4.2.1 Vanadium (V)

A strong carbide stabilizer. It increases wear resistance, hardness, and hot strength by forming very hard vanadium carbides (≈ HV 1500). Used in small amounts (0.10–0.35%) in diesel cylinder liners and wear parts [1–3].

4.2.2 *Titanium (Ti)*

Acts as a graphitizer in small doses (0.10–0.20%) and strongly refines graphite, promoting the undercooled form (Type D). Used to control nitrogen and to avoid cracking defects or compacted graphite. However, it interferes with nodularization in ductile irons [1–3].

4.2.3 *Boron (B)*

A powerful carbide stabilizer. Used in very small amounts (up to 0.002%) to control the annealability of malleable iron and can increase hardenability [1–3].

4.2.4 *Tin (Sn)*

A powerful pearlite stabilizer, even at very low concentrations (up to 0.15%). Used to eliminate free ferrite, which raises strength and hardness without forming massive carbides, improving resistance to sliding wear [1–3].

4.2.5 *Antimony (Sb), Arsenic (As), Bismuth (Bi), Lead (Pb), Tellurium (Te)*

These trace elements, often considered impurities, have very potent—and generally negative—effects.

- **Antimony (Sb)**: Stabilizes pearlite but reduces strength and toughness, and can cause cracking.
- **Arsenic (As)**: Stabilizes pearlite and can improve hot strength, but interferes with nodularization.
- **Bismuth (Bi)**: Can cause Widmanstätten graphite and is detrimental to nodularization, though it is used to control chill in malleable iron.
- **Lead (Pb)**: Highly detrimental; even traces can produce Widmanstätten graphite, which drastically reduces strength.
- **Tellurium (Te)**: A very powerful promoter of chill and can cause mesh (network) graphite, with severe loss of strength [1–3].

4.2.6 *Magnesium (Mg) and Cerium (Ce)*

These two elements are the most important nodulizing/spheroidizing agents, responsible for producing ductile iron.

- **Magnesium (Mg)**: The most commonly used agent for transforming lamellar graphite into spheroidal graphite. Control is critical: too low a residual leads to imperfect graphite (compacted or vermicular), while excess promotes carbide formation.

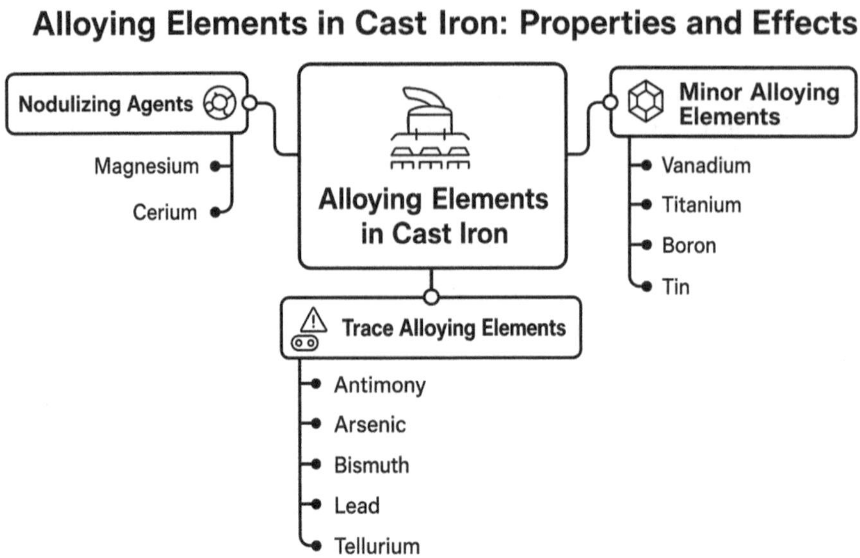

Fig. 4.2 Classification scheme of alloying elements in cast irons, distinguishing between nodulizing agents (Mg, Ce), minor alloying elements (V, Ti, B, Sn), and trace elements (Sb, As, Bi, Pb, Te), and highlighting their functional role in controlling graphite morphology and matrix behavior

- **Cerium (Ce)**: Also a nodulizer and often used together with magnesium to neutralize the harmful effects of trace elements such as lead, bismuth, titanium, and antimony, which interfere with spheroidal graphite formation [1–3].

The addition of alloying elements thus allows cast iron to be transformed into a highly specialized engineering material. Through precise control of composition, it is possible to design alloys with tailored microstructures—from reinforced pearlitic matrices to fully austenitic structures or those filled with high-hardness carbides. Each element is a tool for the metallurgist, enabling optimization of the material for superior performance in applications ranging from high-strength automotive components to parts for the most demanding chemical, mining, and energy industries [1–4] (Fig. 4.2).

4.3 Synergies Between Alloying Elements

4.3.1 The Classic Balance Between Graphitizers and Carbide Formers: Nickel and Chromium

One of the most fundamental and widely used interactions is that between nickel (Ni), a graphitizer, and chromium (Cr), a powerful carbide stabilizer. This counterbalancing synergy underpins several high-performance cast irons.

- In Ni-Hard irons (martensitic white irons): The goal is an extremely hard martensitic matrix without any graphite. Nickel (3.3–5%) is added to suppress the austenite-to-pearlite transformation during cooling, allowing martensite to form. However, nickel also graphitizes, which could cause graphite to appear and ruin the white-iron structure. Chromium (1.4–4%) is therefore added to counter the graphitizing effect of nickel and ensure fully white solidification with stable carbides. The synergy is clear: nickel provides the hardenability needed for a martensitic matrix, while chromium ensures the absence of graphite—indispensable for this type of iron.

- In Ni-Tensil irons (pearlitic gray irons): The aim is a high-strength pearlitic matrix without the brittleness of massive carbides. Here, nickel prevents chill (carbide) in thin sections, refines pearlite, and increases matrix strength. A small addition of chromium is used to stabilize that pearlite, avoid soft ferrite, and raise hardness. Nickel and chromium act together to produce a dense, strong structure in thick sections without sacrificing machinability in thin ones [1–5].

4.3.2 *Combining for Acicular (Bainitic) Structures: Nickel–Molybdenum and Copper–Molybdenum*

To obtain cast irons with a bainitic matrix directly as-cast (so-called acicular irons), the isothermal transformation (TTT) curves must be manipulated to "dodge" pearlite formation. This is achieved via a powerful synergy between nickel (or copper) and molybdenum (Mo).

- Mechanism of the synergy: Nickel (or copper) shifts the entire TTT curve to the right, i.e., delays the start of all transformations (pearlitic and bainitic). Molybdenum is particularly effective at delaying the pearlite "nose." Together, they create sufficient separation between the pearlitic and bainitic transformation curves so that the part's cooling curve in the mold passes through the bainitic field, yielding the desired microstructure. In addition, molybdenum imparts very high toughness to these irons. The nickel and molybdenum contents must be precisely adjusted according to section thickness to control cooling rate [1–5].

4.3.3 *Boosting Hardenability in Thick Sections: Chromium and Molybdenum*

In high-chromium white irons (ASTM A532 Class II and III), the goal is to ensure a hard, strong matrix in large-section parts that cool slowly. Although chromium contributes hardenability, it may be insufficient in very thick sections.

- Synergy for uniform hardness: Chromium (11–30%) is responsible for forming hard M_7C_3 carbides and providing corrosion and heat resistance. To ensure the matrix transforms fully to martensite rather than soft pearlite—especially in the core of thick parts—molybdenum (up to 3%) is added. Molybdenum is a very effective pearlite suppressor that increases hardenability without excessively stabilizing retained austenite, an undesirable side effect that can occur with excess nickel or manganese. Together, Cr and Mo ensure a microstructure of hard carbides in a fully martensitic matrix, even in bulky castings. Nickel and copper are also used for this purpose [1–5].

4.3.4 Neutralizing Impurities and Controlling the Matrix: Manganese and Sulfur

While not a synergy that directly improves properties, the interaction between manganese (Mn) and sulfur (S) is fundamental in the production of nearly all cast irons.

- Corrective synergy: If uncontrolled, sulfur forms iron sulfide (FeS), a low-melting compound that segregates at grain boundaries and causes hot shortness. Manganese's primary role is to combine preferentially with sulfur to form manganese sulfide (MnS), which is far less harmful and acts as a nucleus for lamellar graphite. Only manganese in excess—once all sulfur is neutralized—acts as an alloying element that stabilizes and refines pearlite, increasing hardness [1–5].

4.3.5 Creating Austenitic Structures at Lower Cost: Nickel and Manganese

To obtain a fully austenitic matrix at room temperature, high concentrations of gamma-formers are required. Nickel is the most effective—but also the most expensive.

- Economic and metallurgical synergy: In Nomag irons, designed to be non-magnetic and economical, a synergy between nickel (~10%) and manganese (~5%) is used. Both elements stabilize austenite, and their combined effect achieves a fully austenitic matrix with a lower—and cheaper—total alloy content than using nickel alone. The trade-off: manganese does not provide the corrosion resistance that nickel does, so these alloys are not suitable for corrosive environments [1–5].

4.3.6 Improved Oxidation Resistance: Silicon and Aluminum

Both silicon (Si) and aluminum (Al) dramatically improve oxidation resistance at high temperatures by forming protective surface oxide layers (SiO_2 and Al_2O_3, respectively). When used together, their effect is enhanced.

- Synergy in surface protection: A specific alloy with 6% Al and 2% Si, plus 2% Mo, shows "excellent" oxidation resistance at 925 °C. Silicon and aluminum together form an even denser, more adherent oxide layer than either would alone—especially valuable in heat-resistant ferritic irons [1–5].

4.3.7 Other Relevant Interactions

- Phosphorus with Cr, V, or Mo: Phosphorus (P) forms a hard eutectic called steadite. When carbiding elements such as chromium, vanadium, or molybdenum are also present, phosphorus can form a ternary phosphide–carbide eutectic that is extremely hard (≈780–800 HV), greatly increasing wear resistance—but also brittleness and machining difficulty.
- Cerium with Subversive Elements (Pb, Bi, Ti, Sb): In ductile-iron production, trace elements like lead, bismuth, titanium, and antimony interfere with magnesium's nodulizing action, producing degenerate graphite. Cerium (Ce) is added to neutralize these harmful effects, ensuring proper spheroidal graphite formation [1–5] (Fig. 4.3).

In conclusion, the metallurgy of alloyed cast irons is a complex balancing act. Understanding these synergies enables engineers and foundrymen to design bespoke materials in which the combination of elements produces results that go far beyond the simple sum of their individual effects—achieving optimal performance for the most demanding applications [1–5] (Table 4.1).

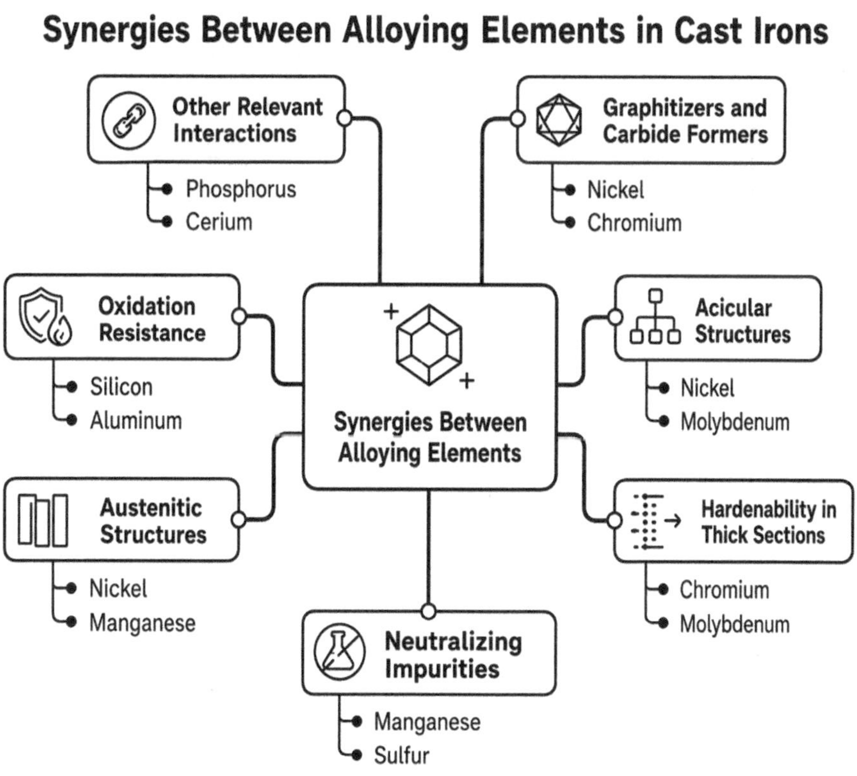

Fig. 4.3 Integrated schematic of the principal synergies between alloying elements in cast irons, including graphitizers vs. carbide formers (Ni–Cr), bainitic structure formation (Ni/Cu–Mo), hardenability in thick sections (Cr–Mo), oxidation resistance (Si–Al), austenitic stabilization (Ni–Mn), and impurity neutralization (Mn–S, Ce with trace elements)

Table 4.1 Minor and trace alloying elements—table (literal translation)

Element	Graphitization/carbides	Effect on matrix/TTT	Properties driven	Risks/limits	Notes/typical uses
C	More C ⇒ more graphite (stable) or more cementite (metastable) depending on system. Lowers Ms if dissolved in γ	Controls phase fractions; hardens martensite via C in solution	+Castability; in white irons +hardness/wear	Excess in grays ↓Rm and E (more graphite)	Key tuning variable per application
Si	Strong graphitizer; prevents chill	Alpha-former: stabilizes ferrite; raises A1; helps obtain stable ferritic matrices (Silal)	+High-T oxidation (SiO_2 layer)	In white irons must be limited to avoid graphitizing	Key in Silal and Si–Al variants
Mn	Mild graphitizer only after neutralizing S; promotes pearlite (in excess)	Increases hardenability; stabilizes pearlite; with Ni can stabilize γ	+Hardness via pearlite; fixes S forming MnS	Excess ⇒ retained austenite; worsens corrosion if replacing Ni	Soaks up S (rule ≈ %Mn ≥ 1.7% S + 0.3). Basis of Nomag (Ni + Mn)
Ni	Moderate graphitizer (≈⅓ of Si)	Gamma-former: refines pearlite; at high % suppresses pearlite (Ni-Hard); >13% ⇒ stable γ (Ni-Resist/ Nomag). Shifts TTT	+Strength, hardness, and toughness (incl. cryogenic); ↓μr; enables low CTE (Minvar)	Excess in Ni-Hard ⇒ high γR	Core in Ni-Tensil, Ni-Hard, Ni-Resist, Minvar, Nomag
Cr	Strong carbide former: M_3C (Cr < 10–12%), M_7C_3 (Cr > 10–12%); inhibits graphite	↑Hardenability; at very high Cr (>28–30%) stabilizes ferrite	+Hardness, +wear, +hot strength and + corrosion; improves impact in Ni-Hard 4	Limited toughness in very high-Cr ferritics	Pillar of Hi-Cr and >28% Cr (abrasion service + 800–900 °C)
Mo	Moderate carbide former; very effectively suppresses pearlite; partitions into carbides and matrix	+Hardenability (thick sections) without over-stabilizing γ; separates pearlite vs bainite (Ni/ Cu + Mo synergy)	+Strength, +hardness, +toughness (acicular); +creep; improves HCl resistance in Durichlor	Poor balance can induce embrittlement if combined improperly	Key in aciculars, Hi-Cr (Class II), Ni-Hard and Si-Mo

Cu	Mild graphitizer; pearlite promoter; at high % stabilizes γ	Shifts TTT (↑hardenability); refines pearlite	+Rm, +HB, +wear; +castability; +atmospheric corrosion	Hinders nodularization (strict control in ductile)	Partial economical substitute for Ni; present in some Ni-Resist
Al	Graphitizer; but 7–18% stabilizes cementite (white); 20–24% graphitizes again (Chugal)	Alpha-former (ferrite); with Si ⇒ very protective Al_2O_3/SiO_2 layer	Extreme + high-T oxidation (4–6% Al or 20–24% Al)	Reactive in melting (defects); cold brittleness	Si–Al (4–6% Al) for furnaces; Chugal (20–24% Al) for maximum oxidation resistance
P	Does not graphitize; forms steadite (phosphide eutectic) and with Cr/V/Mo ternary phosphide–carbide eutectics (780–800 HV)	Hardens interdendritic network; ↑brittleness	+Fluidity; +sliding/scratch wear	↑Brittleness, ↓impact; kept low in high-toughness/ high-T alloys	Useful in automotive grays/ thin parts; limit in Ni-Resist
S	Harmful carbide former; forms FeS if Mn absent	Hot shortness from FeS; Mn neutralizes → MnS	–	↓Rm and ductility; worsens abrasion; ↑melt viscosity	Tight control; adjust Mn/S
Ti	Modifier: favors fine/ undercool graphite (in Silal) with low C	–	Improves toughness by refining graphite (if transformed to nodular with Mg)	Excess may induce unwanted nodules/ compounds	Used as refiner in high-Si
Ce	"Cleans" subversive traces (Pb, Bi, Ti, Sb) so Mg nodularizes properly	–	Ensures correct spheroidal graphite	Critical dosing; excess degrades	Corrector in ductile when impurities present
V/W	Strong carbide formers; promote hard carbides	Harden the matrix; raise hot strength	+Wear/red hardness; T stability	↑Brittleness and machining difficulty	Additives for severe abrasive service

References

1. ASM Handbook Committee. (1990). *ASM handbook, volume 1: Properties and selection: Irons, steels, and high-performance alloys*. ASM International.
2. Davis, J. R. (Ed.). (1996). *ASM specialty handbook: Cast irons*. ASM International.
3. Stefanescu, D. M. (Ed.). (2017). *ASM handbook, volume 1A: Cast iron science and technology*. ASM International.
4. Laird, G., Gundlach, R., & Röhrig, K. (2000). *Abrasion-resistant cast iron handbook*. American Foundry Society.
5. Fashu, S., Xie, Y., Wang, J., & Xiao, W. (2023). Development and performance of high chromium white cast irons (HCWCIs) for wear/corrosive environments: A critical review. *Metals, 13*(11), 1831. https://doi.org/10.3390/met13111831

Chapter 5
Alloyed Cast Irons with Free Graphite

Abstract Alloyed cast irons with free graphite encompass a wide family of gray, ductile and austenitic irons in which carbon precipitates as graphite rather than remaining combined in carbides (ASM Handbook Committee, ASM handbook, volume 1: properties and selection: irons, steels, and high-performance alloys, ASM International, Materials Park, OH, 1990; Davis, ed., ASM specialty handbook: cast irons, ASM International, Materials Park, OH, 1996). Through appropriate alloying with elements such as nickel, chromium, molybdenum, silicon and aluminum, these irons can be tailored to offer high strength pearlitic grades, bainitic or acicular structures, self-hardening martensitic matrices, and a variety of ferritic and austenitic irons for demanding wear-, corrosion- and heat-resistant service (ASM Handbook Committee, ASM handbook, volume 1: properties and selection: irons, steels, and high-performance alloys, ASM International, Materials Park, OH, 1990; Davis, ed., ASM specialty handbook: cast irons, ASM International, Materials Park, OH, 1996; Stefanescu, ed., ASM handbook, volume 1A: cast iron science and technology, ASM International, Materials Park, OH, 2017; Laird et al., Abrasion-resistant cast iron handbook, American Foundry Society, Des Plaines, IL, 2000). Classical examples include Ni-Tensil pearlitic gray iron, bainitic cast irons, self-hardening martensitic irons, high-silicon heat-resistant grades such as Silal, aluminum–silicon ferritic irons, and specialized alloys like Chugal, Duriron, Ni-Resist, Nicrosilal, Minvar and Nomag (ASM Handbook Committee, ASM handbook, volume 1: properties and selection: irons, steels, and high-performance alloys, ASM International, Materials Park, OH, 1990; Davis, ed., ASM specialty handbook: cast irons, ASM International, Materials Park, OH, 1996; Stefanescu, ed., ASM handbook, volume 1A: cast iron science and technology, ASM International, Materials Park, OH, 2017; Laird et al., Abrasion-resistant cast iron handbook, American Foundry Society, Des Plaines, IL, 2000; Davis, ed., Nickel, cobalt, and their alloys, ASM International, Materials Park, OH, 2000). In all cases, the synergy between graphite morphology, matrix type and alloying level governs the balance between strength, toughness, wear resistance, thermal stability and corrosion behavior, providing a large design space for engineering applications where castability, cost and service conditions must be simultaneously optimized (ASM Handbook Committee, ASM handbook,

M. González Fernández de Castro et al., *Alloyed Cast Iron*, SpringerBriefs in Materials, https://doi.org/10.1007/978-3-032-27303-1_5

volume 1: properties and selection: irons, steels, and high-performance alloys, ASM International, Materials Park, OH, 1990; Davis, ed., ASM specialty handbook: cast irons, ASM International, Materials Park, OH, 1996; Stefanescu, ed., ASM handbook, volume 1A: cast iron science and technology, ASM International, Materials Park, OH, 2017; Laird et al., Abrasion-resistant cast iron handbook, American Foundry Society, Des Plaines, IL, 2000; Davis, ed., Nickel, cobalt, and their alloys, ASM International, Materials Park, OH, 2000).

Keywords Alloyed cast irons with free graphite · Ni-Tensil · Bainitic cast iron · Martensitic cast iron · Silal · Chugal · Duriron · Ni-Resist · Nicrosilal · Minvar · Nomag · Heat- and wear-resistant irons

5.1 Low-Alloy Pearlitic Gray Irons (Ni-Tensil)

5.1.1 Definition, Composition, and Classification

Low-alloy pearlitic cast iron, commonly known by its trade name "Ni-Tensil," is a type of gray cast iron of high mechanical strength that is distinguished by the controlled addition of elements such as nickel, chromium, molybdenum, and copper. It is classified within low- to medium-alloy cast irons, since the total sum of these elements usually does not exceed 5 wt%. Its development, patented in the 1920s by Mond Nickel Co., represented a significant advance by seeking to overcome the limitations of common gray cast irons, achieving for the first time tensile strengths above 345 MPa (50,000 psi).

The typical chemical composition of a Ni-Tensil casting falls within the following ranges:

- Carbon (C): 3.0–3.5%
- Silicon (Si): 1.5–2.9%
- Manganese (Mn): 0.6–0.9%
- Nickel (Ni): 0.5–3.5%
- Chromium (Cr): 0.3–1.0%
- Molybdenum (Mo): 0.2–1.0%
- Copper (Cu): 0.2–1.0%

Small amounts of titanium (Ti) or vanadium (V) can occasionally be added to refine both the matrix grain size and the graphite morphology [1–5].

5.1.2 Microstructure and Role of the Alloying Elements

The characteristic microstructure of a Ni-Tensil consists of lamellar graphite (generally Type A) dispersed in a predominantly pearlitic matrix. The aim of the alloying is precisely to ensure the formation of this fine, strong pearlitic matrix, which offers greater wear resistance compared with ferritic matrices. The appearance of free ferrite, which would soften the material, or of massive carbides, which would embrittle it, is to be avoided. Metallurgically, it is interesting to note that an insufficient amount of nickel in a Ni-Hard casting (which seeks to be martensitic) would result in the formation of this pearlitic structure, which underscores the importance of precise composition control to achieve the desired matrix in each type of alloy.

The synergistic combination of the elements is key to this balance:

- Nickel (Ni): It is the principal alloying element and the one that gives the material its name. It acts as a moderate graphitizer (approximately one third of the effectiveness of silicon), which helps to prevent white solidification (chill) in thin sections and corners, improving machinability. In addition, it refines the pearlite and the grain size, increasing the hardness and strength of the matrix. Additions of 0.5% to 3.5% make it possible to obtain strengths of 35 to 50 kg/mm^2.
- Copper (Cu): A mild graphitizer, but in practice it acts above all as a pearlite promoter: it helps eliminate free ferrite, refines pearlite, and raises the strength of the matrix (an effect similar to that of nickel). In combination with molybdenum it improves microstructural control (hardenability/refinement). Up to ~1% it usually improves corrosion resistance in industrial atmospheres and weak acids, and it also provides anti-friction qualities.
- Chromium (Cr): A carbide stabilizer that counteracts the graphitizing effect of nickel and silicon. It increases the hardness of the matrix, refines pearlite, and can form small amounts of free chromium carbides that improve wear resistance.
- Molybdenum (Mo): Increases the strength and hardness of the matrix, strongly refines the graphite, and depresses the transformation temperature of pearlite, which contributes to a finer, stronger structure [1–5].

5.1.3 Phosphorous Variant and Steadite

A widely used variant of this cast iron is the phosphorous type, which contains between 0.5% and 1.5% phosphorus (P). Phosphorus forms a low-melting eutectic (approximately 930 °C) known as steadite, which solidifies in the interdendritic zones. This low-melting microconstituent confers excellent castability on the iron, making it ideal for parts of complex shapes or thin walls. Steadite is a very hard constituent, with values between 400 and 600 HB (530–600 HV), and at room temperature it is composed of a binary eutectic of small ferritic globules in a matrix of iron phosphide (Fe_3P), with a composition of 61% Fe_3P and 39% ferrite. Its presence in the form of an intercellular network (from 0.4% to 0.7% P) significantly increases the overall hardness (about 30–40 HB for each 1% P) and the resistance to wear and to scratching (scuffing) [1–5] (Fig. 5.1).

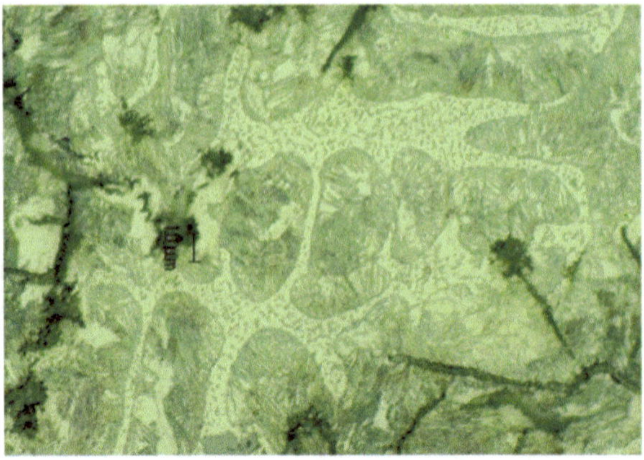

Fig. 5.1 Micrograph of steadite in pearlitic Ni-Tensil cast iron etched with Nital 3%, showing the interdendritic phosphide eutectic network embedded within the pearlitic matrix

5.1.4 Mechanical Properties

Ni-Tensil cast irons stand out for mechanical properties superior to those of common gray cast irons, and for the possibility of being hardened by heat treatment up to strengths of 900–1000 MPa.

- Tensile Strength: Lies between 250 and 500 MPa in the as-cast state. One of their great advantages is their low sensitivity to section thickness, which makes it possible to obtain uniform properties in parts of variable sections.
- Hardness: The alloyed pearlitic matrix is significantly harder than that of common cast irons; Knoop hardness values of 305 have been measured for the alloyed pearlite, compared with 190 RV/235 Knoop for the unalloyed. The overall Brinell hardness varies between 180 and 300 HB. In the phosphorous variants, the presence of steadite can raise the local hardness up to 400–600 HB.
- Fatigue Strength: The fatigue limit for machined parts is estimated at 35–45% of their minimum tensile strength. As tensile strength increases, the vibration damping capacity tends to decrease.
- Toughness: Although they are brittle materials, their toughness can be evaluated indirectly through the transverse strength test, where the product of the strength and the deflection is a measure of the energy absorbed before fracture [1–5].

5.1.5 Physical, Electrical, and Magnetic Properties

- Density: The typical density is approximately 7.2 g/cm^3, although it can vary between 7.05 and 7.30 g/cm^3. The specific gravity of its constituents is 7.78 for pearlite and 7.32 for steadite.

- Thermal Conductivity: It exhibits good conductivity, in the range of 40–50 W/m K, thanks to the presence of graphite. It decreases with temperature, ranging from 43.5–65.7 W/m K at 100 °C to 36.0–40.9 W/m K at 500 °C.
- Coefficient of Thermal Expansion: Lies at about 11–12.5 × 10^{-6} °C^{-1}. This moderate expansion, lower than that of austenitic Ni-Resist cast irons, gives it better thermal shock resistance.
- Electrical Properties: Electrical resistivity lies between 65 and 80 μΩ cm. Phosphorus above 0.2% tends to increase resistivity.
- Magnetic Properties: They are ferromagnetic due to their pearlitic matrix. Nickel and copper tend to decrease permeability and increase hysteresis losses.
- Service Temperature: The recommended maximum service temperature is approximately 400 °C [1–5].

5.1.6 Industrial Applications

Thanks to their combination of high strength, good wear resistance, machinability, and castability, Ni-Tensil cast irons are used in a wide range of demanding engineering components.

- Automotive and Machinery Industry: Used to manufacture cylinder liners, brake drums and shoes, piston rings, crankshafts, camshafts, and machine beds. They are suitable for conditions of poor lubrication where the presence of free ferrite would be detrimental.
- Railway and Mining Industry: They also find use in railroad wheels and brake shoes, as well as in mill liners and hammers for mining equipment.
- Tooling: Used in dies for stamping where good surface wear resistance is required.
- Phosphorous Applications: The exceptional fluidity of the phosphorous variant makes it ideal for wear parts such as liners, rings, and brake shoes, as well as for the manufacture of thin-walled parts and intricate designs, such as radiators and artistic castings [1–5] (Fig. 5.2).

Fig. 5.2 Property summary diagram of Ni-Tensil cast iron, correlating chemical composition, lamellar graphite in pearlitic matrix, tensile strength (250–500 MPa), hardness (180–300 HB), density, thermal conductivity, electrical resistivity, magnetic behavior, and typical industrial applications

Ni-Tensil Cast Iron Properties

Ni-Tensil	
Composition	Ni, Cr, Mo, Cu added
Microstructure	Lamellar graphite in pearlitic matrix
Tensile Strength	250–500 MPa
Hardness	180–300 HB
Density	7.05–7.30 g/cm^3
Thermal Conductivity	40–50 W/m·K
Electrical Resistivity	65–80 μΩ·cm
Magnetic Properties	Ferromagnetic
Applications	Cylinder liners, brake drums

5.2 Bainitic or Acicular Cast Iron (with Free Graphite)

5.2.1 *Definition, Nomenclature, and Microstructure*

Bainitic cast iron with free graphite, commonly called acicular cast iron, is a type of high-alloy cast iron characterized by a microstructure whose matrix is predominantly bainitic, obtained directly in the as-cast state. The term "acicular" refers to the needle-like morphology (aciculae) of the bainite that makes up the matrix. This microstructure is achieved by careful control of chemical composition and cooling rate, without the need for subsequent quench heat treatments. In some variants, such as so-called acicular "NI-TENSIL," the structure may consist of graphite flakes on a bainitic matrix with isolated zones of troostite [1–5].

The metallurgical key to obtaining this structure lies in manipulating the isothermal transformation (TTT) curves. The addition of specific alloying elements shifts the pearlitic region of the TTT curve to the right, i.e., to longer times, allowing the casting's cooling curve in the mold to "dodge" the pearlitic nose and intersect the bainitic transformation field. To maximize mechanical strength and, above all, toughness, it is essential that the final microstructure be free of pearlite [1–5].

5.2.2 *Chemical Composition and Role of Alloying Elements*

Obtaining a bainitic matrix in the as-cast state requires a very precise chemical formulation, which must be adjusted according to section thickness to control cooling rate. The typical composition of an acicular cast iron includes:

- Carbon (C): 2.9–3.2%
- Silicon (Si): ~2.0%
- Manganese (Mn): ~1.0%
- Nickel (Ni): 1.5–2.2%
- Molybdenum (Mo): 0.3–0.95% [1–5]

Nickel and molybdenum are the fundamental alloying elements for this process. Nickel shifts both the pearlitic and the bainitic transformation fields to the right (to longer times). Molybdenum, for its part, is particularly effective at delaying the pearlitic transformation, chiefly shifting the Ar'' transformation point. Molybdenum is of special interest in these cast irons, since it imparts very high toughness to the alloy. The nickel content is adjusted according to section thickness: parts under 38 mm (1.5 in.) may require 1.0–1.4% Ni, whereas sections of 51–102 mm (2–4 in.) may need up to 2.5–3.0% Ni to ensure bainite formation throughout the thickness [1–5].

5.2.3 Mechanical and Physical Properties

Acicular cast irons stand out for their excellent combination of high strength and good toughness.

- Tensile Strength: Ranges between 400 and 700 MPa. The acicular "NI-TENSIL" variant reaches 40–60 kg/mm^2 (392–588 MPa).
- Hardness and Machinability: Hardness can reach up to 320 HB and still be machinable, thanks to the absence of free cementite in the microstructure. They offer acceptable machinability, provided they are free of pearlite.
- Toughness: They possess very high toughness. A commercial example shows an impact strength of 44 ft lb (60 J) on an unnotched 0.798-in. specimen.
- Physical Properties: For a commercial acicular cast iron, the following properties have been measured:
  - Density: 7.32 g/cm^3.
  - Electrical Resistivity: 66.13 μΩ cm.
  - Coefficient of Thermal Expansion (×10^{-6}/°C): 14.4 (20–100 °C), 14.5 (20–200 °C), 14.8 (20–300 °C), and 15.2 (20–400 °C) [1–5].

5.2.4 Heat Treatments

Although their main feature is obtaining the bainitic matrix in the as-cast condition, subsequent heat treatments can be applied to further optimize properties. A treatment at 350 °C for 7 h, followed by furnace cooling, can improve performance [1–5].

5.2.5 Industrial Applications

Thanks to their high tensile strength, good toughness, and wear resistance, combined with acceptable machinability, acicular cast irons are used in components subjected to high mechanical loads. The most common applications include:

- Engine components: Crankshafts, camshafts, and gears.
- Tooling and dies: Used for hot-forming dies, such as those used to mold shells, where the operating temperature does not exceed 400–460 °C. They are also valuable as die blocks for hot or cold forging [1–5].

5.2.6 Distinction from Austempered Ductile Iron (ADI)

It is crucial not to confuse as-cast acicular cast iron with Austempered Ductile Iron (ADI), although both have a bainitic matrix (or ausferritic in the case of ADI). The fundamental differences are [1–5]:

Comparison of Acicular Cast Iron and ADI

Characteristic	Graphite Morphology	Matrix Formation	Production Process	Tensile Strength
Acicular Cast Iron	Flake graphite	As-cast bainitic matrix	Mold cooling with Ni–Mo alloying	400–700 MPa
Austempered Ductile Iron (ADI)	Spheroidal graphite	Austempered ausferritic matrix	Austempering heat treatment	950–1600 MPa

Fig. 5.3 Comparison chart between as-cast acicular (bainitic) cast iron and austempered ductile iron (ADI), contrasting graphite morphology, matrix formation route (as-cast bainite vs. austempered ausferrite), production process, and tensile strength range

- *Graphite Morphology*: Conventional acicular cast iron has flake graphite, whereas ADI is produced from ductile iron with spheroidal graphite [1–5].
- *Production Process*: The bainitic matrix of acicular cast iron is obtained directly during mold cooling (as-cast) thanks to Ni–Mo alloying. In contrast, ADI's matrix is obtained via a specific isothermal heat treatment called austempering, applied to the part after solidification [1–5].

The austempering process for ADI involves heating the ductile iron casting to an austenitizing temperature (850–900 °C), rapidly cooling it to an intermediate temperature (250–400 °C) in a salt bath, holding isothermally until transformation to ausferrite is complete (approximately 45 min), and finally air cooling. This controlled process allows ADIs to reach mechanical properties even higher than acicular cast irons (tensile strength of 950–1600 MPa) with significantly greater ductility [1–5] (Fig. 5.3).

5.3 Self-Hardening Martensitic Cast Iron (with Free Graphite)

5.3.1 Definition, Microstructure, and Metallurgy

Martensitic cast iron with free graphite is a specialized iron alloy, often classified as a self-hardening gray cast iron. Its microstructure is characterized by a combination of flake graphite, a predominantly martensitic matrix, and variable amounts of retained austenite (γR). The defining feature of these alloys is their self-hardening property, meaning their matrix transforms to martensite directly during normal cooling in the mold, without the need for rapid quenching or subsequent hardening [1–5] (Fig. 5.4).

Metallurgically, its composition is designed to achieve a dual objective. On the one hand, the silicon content is high enough to promote stable solidification, ensuring that carbon precipitates as free graphite and thus preventing the formation of a

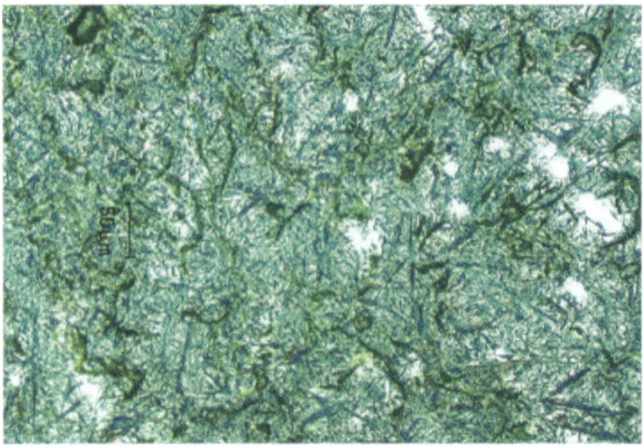

Fig. 5.4 Micrograph of self-hardening martensitic cast iron etched with Nital 3%, showing flake graphite dispersed in a predominantly martensitic matrix with retained austenite

white-iron structure. On the other hand, a high nickel content slows the transformation rate of austenite to pearlite, allowing the mold cooling to avoid the pearlitic "nose" and for the transformation to occur directly to martensite at lower temperatures. This presence of free graphite is the main difference from martensitic white irons such as Ni-Hard. Although the presence of graphite reduces the overall hardness of the material compared with what a pure martensitic matrix would have (whose intrinsic hardness can be 770–800 HV or 61.5–62 HRC), it is precisely this graphite that confers a certain degree of machinability to the alloy [1–5].

5.3.2 Chemical Composition and Role of Alloying Elements

The chemical composition of these cast irons is carefully tuned to achieve the self-hardening microstructure. Typical ranges are as follows:

- Carbon (C): 2.90–3.30%
- Silicon (Si): 1.50–2.00%
- Manganese (Mn): 0.7–1.0%
- Chromium (Cr): 0.6–1.0%
- Nickel (Ni): 4.0–5.0% [1–5]

Nickel is the key alloying element, since its high concentration depresses the pearlitic transformation temperature and confers the self-hardening property. Chromium, in moderate contents of 0.5–0.8%, is considered beneficial for stabilizing the structure. A technical source cites a specific example with a composition of 3.3% C, 1.05% Si, 0.71% Mn, 4.95% Ni, and 0.49% Cr [1–5].

5.3.3 Mechanical and Physical Properties

These cast irons are characterized by a combination of high hardness and inherent brittleness, which defines their application field.

- Hardness: Their Brinell hardness lies in the range of 400–500 HB. This high hardness is the direct result of the martensitic matrix. A specific commercial example showed a hardness of 404 HB.
- Tensile Strength: Sources cite a tensile strength of 19 ton/in.^{2} (approximately 293 N/mm^2) for a particular composition.
- Brittleness and Toughness: They are considered very brittle materials. This brittleness is the main reason why their use in the as-cast state is not common. Very brittle when cold; improves above ~250 °C. Use is recommended at elevated temperatures, with caution regarding thermal shock when cold.
- Machinability: They are difficult to machine. However, unlike white irons containing massive free carbides, these alloys can be machined using appropriate speeds and feeds.
- Service Temperature: The maximum recommended service temperature is around 350 °C [1–5].

5.3.4 *Heat Treatments and Alternative Processing Routes*

Although these cast irons are self-hardening, their use in the as-cast state is uncommon due to their brittleness. To improve their properties, subsequent heat treatments are usually applied:

- Tempering: The most common treatment. Tempering at temperatures between 200 and 300 °C slightly increases toughness without an appreciable loss of hardness. A more specific treatment of 5 h at 230 °C has been shown effective at eliminating most of the retained austenite in the microstructure.
- Alternative Production Route: There is an interesting compositional variant, with 6.5% Ni and 3.75% Mn, that offers a different processing route. In the as-cast state, this alloy presents an austenitic matrix with a low hardness of 160 HB. This allows the part to be machined in a relatively soft state. Subsequently, a heat treatment at 600 °C for 15 hours transforms the matrix into a hard martensitic structure with a hardness of 410–420 HB [1–5].

5.3.5 *Applications and Comparison with Alternatives*

Due to their profile of high hardness and wear resistance combined with low toughness, their applications are very specific. They are used in parts that operate under wear without being subjected to heavy impacts and that, ideally, can be used directly without the need for machining [1–5].

These cast irons compete directly with gray cast iron parts that have been surface-hardened by flame or induction, since the latter have a very similar surface structure and properties (an outer layer of martensite and graphite). The main difference is that self-hardening cast irons acquire their hardness throughout the entire volume of

Comparison of Martensitic Cast Iron Alloys

Characteristic	Matrix	Graphite	Hardness	Machinability	Key Alloying Elements	Heat Treatment	Applications
Self-Hardening Martensitic Cast Iron	Predominantly martensitic	Flake graphite	400–500 HB	Difficult	Nickel	Tempering	Wear parts
Ni-Hard Cast Iron	Martensitic	No free graphite	Higher than 500 HB	Very difficult	Nickel and Chromium	Not applicable	Abrasive wear parts

Fig. 5.5 Comparative table of martensitic cast iron alloys, contrasting self-hardening martensitic cast iron (with flake graphite) and Ni-Hard white cast iron (without free graphite) in terms of matrix, graphite presence, hardness, machinability, key alloying elements, heat treatment requirements, and typical applications

the part during casting, whereas surface-hardened parts achieve it only in the outer layer through a subsequent treatment [1–5].

It is important not to confuse them with Ni-Hard cast irons, since the presence of free graphite clearly distinguishes them, and there is no specific standard that covers them [1–5] (Fig. 5.5).

5.4 High-Silicon Ferritic Cast Iron (Silal)

5.4.1 Introduction and Historical Context

Silal cast iron is a heat-resistant iron alloy developed in the 1930s by the British Cast Iron Research Association (BCIRA). It is classified as a high-silicon ferritic cast iron (≈5.5–7 wt% Si) and is one of the most economical options for applications requiring high oxidation resistance at elevated temperatures. Its hallmark is a stable ferritic microstructure that, thanks to the high silicon content, offers excellent resistance to dimensional growth and scale formation in continuous or intermittent service up to ≈850–900 °C.

Oxidation-resistance mechanism. The high Si content promotes, during service, the formation of a dense, strongly adherent surface layer rich in SiO_2 and iron silicates (e.g., fayalite). This film drastically reduces oxygen diffusion into the base metal compared with conventional iron oxides, stabilizing the surface and maintaining dimensional integrity of the part [1–5].

5.4.2 Chemical Composition and Role of Alloying Elements

Silal's chemistry is tightly controlled to achieve its specific properties. Typical ranges:

- Carbon (C): <2.3%
- Silicon (Si): 5.5–7.0%

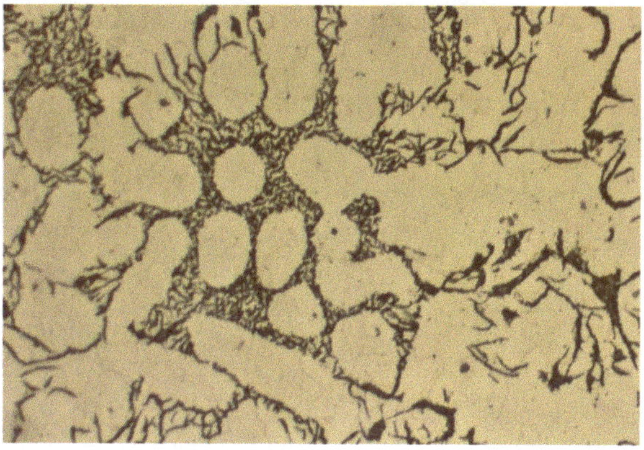

Fig. 5.6 Metallographic image of high-silicon ferritic cast iron (Silal) etched with Nital, showing fine interdendritic graphite within a ferritic matrix stabilized by elevated silicon content

- Manganese (Mn): 0.5–0.8%
- Phosphorus (P): 0.1–0.3%
- Sulfur (S): <0.06%

Occasionally, chromium (Cr) up to ~2% is added to further improve oxidation resistance. To optimize creep strength at elevated temperature, 0.5–2% molybdenum (Mo) may be added [1–5].

5.4.3 Microstructure and Metallurgy

Silal's microstructure consists of free graphite of interdendritic, undercooled Type D morphology dispersed in a fully ferritic matrix. This structure is the direct result of its high silicon concentration. Silicon is a powerful graphitizer, ensuring that carbon solidifies as graphite rather than iron carbide (cementite). In addition, silicon is a strong ferrite stabilizer (alpha-former), meaning it markedly raises the A1 allotropic transformation temperature (the temperature at which ferrite transforms to austenite on heating). In Silal irons this critical temperature is raised to ~970 °C. As a result, the matrix remains ferritic and structurally stable throughout its usual service temperature range, preventing the volume changes associated with phase transformations and conferring excellent resistance to growth.

For best properties, fine, undercooled (Type D) graphite is sought; this can be favored by keeping carbon low and adding titanium. The microstructure—and hence the mechanical properties—can be markedly improved by adding magnesium (Mg), which converts interdendritic graphite to spheroidal graphite, greatly increasing toughness and tensile strength [1–5] (Fig. 5.6).

5.4.4 Oxidation and Heat Resistance

Silal's standout property is its excellent high-temperature behavior. Its oxidation and growth resistance stem mainly from two metallurgical factors [1–5]:

- *Formation of a protective oxide layer*: On exposure to high temperature, silicon at the surface reacts with oxygen to form a dense, adherent, stable surface layer of silicon oxide (iron silicate). This layer is far more resistant to oxygen ingress than conventional iron oxide, effectively protecting the underlying metal. Protection improves as silicon increases; above ~4% Si it suffices to prevent significant weight gain [1–5].
- *Stability of the ferritic matrix*: As noted, the matrix remains ferritic up to ~970 °C, avoiding the ferrite ↔ austenite phase cycles that in common gray irons cause volume changes leading to cracking and irreversible dimensional growth. BCIRA tests showed negligible growth below 400 °C, ~0.1% at 800 °C, and 1–1.5% at 900 °C after 64 weeks.

These features allow Silal to operate continuously or intermittently in the 650–850 °C range, though tensile strength drops at very high temperatures [1–5].

5.4.5 Mechanical and Physical Properties

- Strength and Hardness: Silal shows tensile strength (Rm) of 160–278 MPa with interdendritic graphite. If the graphite is spheroidal, strength rises to 200–230 MPa. Brinell hardness is ~140–190 HB.
- Toughness and Brittleness: A key feature at room temperature is brittleness. High silicon in ferrite embrittles it, yielding low toughness and poor thermal-shock resistance in the cold. However, this improves markedly above ~250 °C, where impact resistance can even surpass that of ordinary gray irons.
- Physical Properties:
  - Density: 6.9–7.4 g/cm^3.
  - Linear thermal expansion coefficient: ~12.3 × 10^{-6}/°C (20–100 °C), rising to ~15.4 × 10^{-6}/°C at 500 °C; values near ~18 × 10^{-6}/°C appear above ~600–700 °C.
  - Electrical resistivity: High, ~117–160 μΩ cm.
  - Thermal conductivity: ~37 W/m K at 100 °C (other sources report slightly different values) [1–5].

5.4.6 Production, Heat Treatment, and Machinability

Silal can be produced in cupola furnaces, though electric melting is preferred for better composition control and to avoid hydrogen pickup. While castable, it demands tight control: high silicon favors oxide formation and inclusions and can cause filling issues; hence its industrial use is more limited than common gray irons. To ensure a stable, fully ferritic microstructure, annealing is often recommended—either a full anneal at 900 °C with slow cooling, or a subcritical anneal at 720–790 °C. Unlike very high-silicon irons such as Duriron, Silal is machinable, though carbide tooling is required due to its hardness and brittleness [1–5].

5.4.7 Industrial Applications

Thanks to excellent heat and oxidation resistance at relatively low cost, Silal is used mainly in high-temperature components such as:

- Furnace parts: Grates, ash pans, supports, and other internal parts of industrial furnaces and stoves.
- Engines and Turbines: Gas-turbine components and parts exposed to hot gases.
- Other: Burner nozzles and heat-treatment trays.

In short, Silal is a high-performance cast iron for high-temperature service, offering an economical, effective solution when oxidation and growth resistance are paramount and mechanical loads and room-temperature thermal shocks are not critical. It is distinguished from Duriron by its lower silicon content, making it machinable and tougher, and from Nicrosilal, an austenitic variant that is much more expensive but far superior in toughness and thermal-shock resistance [1–5] (Fig. 5.7).

Silal Cast Iron Properties

Property	Description
Chemical Composition	C < 2.3%; Si: 5.5–7.0%; Mn: 0.5–0.8%; P: 0.1–0.3%; S: <0.06%; Cr: up to ~2%; Mo: 0.5–2%
Microstructure	Free graphite in ferritic matrix
Oxidation Resistance	Dense SiO_2 and iron silicate layer
Heat Resistance	Stable ferritic matrix up to ~970 °C
Strength	Tensile strength: 160–278 MPa (interdendritic), 200–230 MPa (spheroidal)
Hardness	~140–190 HB
Toughness	Brittle at room temperature, improves above ~250 °C
Density	6.9–7.4 g/cm^3
Thermal Expansion	~12.3×10^{-6}/°C (20–100 °C), ~15.4×10^{-6}/°C (500 °C), ~18×10^{-6}/°C (600–700 °C)
Electrical Resistivity	~117–160 μΩ·cm
Thermal Conductivity	~37 W/m·K at 100 °C
Production	Cupola or electric melting, tight control needed
Heat Treatment	Annealing recommended for stable microstructure
Machinability	Machinable with carbide tooling
Applications	Furnace parts, engine components, burner nozzles

Fig. 5.7 Property infographic for Silal cast iron summarizing composition, microstructure (graphite in ferritic matrix), oxidation/heat resistance mechanism, tensile strength and hardness ranges, key physical properties, production/heat-treatment notes, machinability, and applications

5.5 Ferritic Cast Iron with Free Graphite Alloyed with Aluminum and Silicon

5.5.1 Introduction, Classification, and Metallurgical Context

Ferritic cast iron alloyed with aluminum and silicon is a specialized variant of heat-resistant irons, often regarded as a derivative of Silal. Its main feature is the partial substitution of silicon by aluminum to leverage the latter's strong graphitizing effect and, above all, its ability to enhance high-temperature oxidation resistance. These alloys are classified among high-alloy graphitic irons, specifically with ferritic matrix.

Despite attractive properties, commercial production is uncommon. The reason: they are difficult to manufacture because aluminum's high reactivity with oxygen causes major melting and molding issues—slag formation, oxide inclusions, and filling defects ("cold shuts"). In general classifications of Al-bearing cast irons, two main groups are distinguished: one with 1–7% Al and another with 18–25% Al; this alloy clearly falls in the first group [1–5].

5.5.2 Chemical Composition and Role of Alloying Elements

Chemistry is carefully balanced to obtain a ferritic graphite-bearing structure while avoiding excessive brittleness and manufacturing problems. Typical ranges:

- Carbon (C): 2.5–3.5%
- Silicon (Si): 1.6–2.1%
- Aluminum (Al): 4.0–6.0%
- Manganese (Mn): 0.1–0.4%.

Aluminum is the key alloying element, with a complex role. It is a strong graphitizer, favoring graphite over carbides. Its effectiveness varies with content: graphitizing action decreases between 2–4% Al, but in the 4–6% range graphite formation is ensured. Aluminum is also a strong ferrite stabilizer, like silicon, raising the pearlitic transformation temperature and promoting a ferritic matrix stable at high temperature. Molybdenum (Mo) up to ~2% can be added to significantly improve both high-temperature oxidation resistance and creep strength [1–5].

5.5.3 Microstructure

These irons consist of free graphite dispersed in a predominantly ferritic matrix. As in Silal, silicon dissolved in ferrite contributes to brittleness. Graphite morphology can be controlled:

- Flake Graphite: Obtained in the as-cast state without further treatment.

- Spheroidal Graphite: Obtainable via magnesium (Mg) addition, which drastically improves mechanical properties. In Al-bearing ductile irons, a "bull's-eye" structure is common, where graphite nodules are encased by a ferrite envelope [1–5].

5.5.4 *Properties and Service Characteristics*

Designed for superior performance at high temperature.

5.5.4.1 Heat and Oxidation Resistance

This is the standout property. Excellent oxidation and scaling resistance at high temperature stems from aluminum's ability to form a surface layer of aluminum oxide (Al_2O_3) that is dense, adherent, and highly resistant to oxygen penetration.

- Service temperature: Usable in continuous service up to ~950 °C.
- Synergistic effect: Combining Si and Al yields synergy, forming even more protective oxide layers. Adding ~2% Mo to an alloy with 6% Al and 2% Si provides "excellent" oxidation resistance at 925 °C, further improved if the graphite is spheroidal [1–5].

5.5.4.2 Mechanical Properties

Strongly dependent on graphite morphology:

- Tensile Strength (Rm):
  - With flake graphite: 175–225 MPa.
  - With spheroidal graphite: 390–570 MPa, a very significant increase.
- Toughness and Brittleness: These alloys are brittle and have low thermal-shock resistance. Behavior resembles Silal's, though thermal-shock resistance improves above ~250 °C.
- Machinability: Difficult to machine. Although a ferritic matrix might suggest better machinability, material brittleness and hard aluminum oxides complicate the process. Still, they are more machinable than >8% Al variants, which require grinding [1–5].

5.5.4.3 Physical Properties

- Density: Aluminum markedly lowers density in cast irons, so these alloys are lighter than conventional gray irons.
- Thermal Conductivity: Aluminum also markedly reduces thermal conductivity, so these irons will conduct heat less than unalloyed gray irons [1–5].

5.5.5 Industrial Applications

Given their properties, these irons suit components operating at high temperature in oxidizing environments where mechanical loads and impacts are not severe. Typical uses include:

- Hot parts of furnaces and combustion equipment.
- Components where oxidation resistance is the primary requirement and strong mechanical or thermal shocks are not expected.

In summary, Al–Si ferritic cast irons are a niche solution for high-temperature applications, offering oxidation resistance superior to conventional Silal irons. However, adoption is limited by significant casting-process difficulties and inherent brittleness [1–5] (Fig. 5.8).

Ferritic Cast Iron with Al and Si

Characteristic	Flake Graphite	Spheroidal Graphite
Tensile Strength (MPa)	175–225	390–570
Oxidation Resistance	Good	Excellent
Machinability	Difficult	Difficult
Thermal Conductivity	Low	Low
Density	Low	Low
Toughness	Low	Low
Thermal-Shock Resistance	Low	Low

Fig. 5.8 Comparative table for ferritic cast iron alloyed with Al and Si, contrasting flake-graphite versus spheroidal-graphite variants in tensile strength range and qualitative performance (oxidation resistance, machinability, thermal conductivity, density, toughness, and thermal-shock resistance)

5.6 Ferritic Cast Iron with Free Graphite: Chugal (20–24% Al)

5.6.1 Introduction and Classification

Chugal cast iron is a highly specialized ferrous alloy, notable for its exceptional oxidation resistance at extremely high temperatures. It is classified within the group of graphitic irons with a ferritic matrix, but it is distinguished by a very high aluminum content, typically between 20% and 24%, placing it in the high-alloy category (>10%). Unlike other aluminum-alloyed cast irons with lower percentages, Chugal exhibits a microstructure in which virtually all the carbon precipitates as graphite in a matrix that is completely ferritic and stable. Its chief appeal is its unmatched heat resistance, surpassing all other cast irons. However, this exceptional performance comes with significant challenges, such as high brittleness at room temperature and considerable manufacturing difficulties, which limit its applications to very specific niches [1–5].

5.6.2 Chemical Composition and Role of Alloying Elements

The chemical composition of Chugal irons is very particular and centered on maximizing the effect of aluminum. Typical ranges are:

- Carbon (C): ~2%
- Silicon (Si): ~1%
- Aluminum (Al): 20–24%
- They may occasionally contain traces of Manganese (Mn) (~0.26%) and Sulfur (S) (~0.01%) [1–5]

Aluminum is the dominant element and its role is multifaceted:

- Graphitizer: At these concentration levels (up to 25%), aluminum strongly intensifies graphitization, ensuring that all the carbon precipitates in its free form.
- Ferrite Stabilizer: A strong alpha-forming element, guaranteeing a completely ferritic, stable matrix.
- Paramagnetic Effect: When aluminum content exceeds ~20%, alpha iron loses its ferromagnetic properties, making the alloy paramagnetic.
- Oxidation Resistance: Responsible for forming a protective aluminum oxide (Al_2O_3) layer that is extremely stable at high temperatures [1–5].

Use of cerium (Ce) as a modifying agent has also been documented; it considerably improves the alloy's high-temperature performance [1–5].

5.6.3 *Microstructure*

Chugal's microstructure consists of free graphite dispersed in a purely ferritic (alpha-phase) matrix whose structural components are very stable even at high temperatures. Thanks to aluminum's strong graphitizing action in this range, virtually all carbon is present as graphite. The graphite morphology is often globular, especially if the iron has been modified with ferrocerium. However, some sources note the possible presence of complex eutectic carbides of the type Fe_3AlC_x arranged at grain boundaries and within ferrite grains. In the case of cerium modification (in alloys with 19–25% Al), an additional ferromagnetic delta phase may appear, located mainly along the grain boundaries of the alpha phase [1–5].

5.6.4 *Properties and Service Characteristics*

5.6.4.1 Heat, Oxidation, and Corrosion Resistance

This is Chugal's most outstanding property. It offers the greatest oxidation and heat resistance of all cast irons, able to operate continuously at 1000–1100 °C. Its resistance to dry corrosion is extraordinary, and it also exhibits high resistance to sulfuric acid and sulfur-bearing atmospheres. Cerium modification further improves high-temperature behavior [1–5].

5.6.5 *Mechanical Properties*

- Tensile Strength (Rm): At room temperature, it shows surprisingly good strength, 294–343 MPa (30–35 kg/mm^2). However, this strength decreases with temperature, dropping to about 147 MPa (15 kg/mm^2) at 800 °C.
- Brittleness: Its principal drawback is very high brittleness at room temperature, which drastically limits applications. It cannot withstand impact loads or mechanical shocks.
- Hardness: Typically 250–500 HB.
- Machinability: Difficult to machine, possibly requiring grinding for finishing [1–5].

5.6.6 *Magnetic and Physical Properties*

- Magnetism: Due to the high aluminum content (>20%), the matrix's alpha iron loses ferromagnetism, making Chugal paramagnetic. Although cerium modification can introduce a ferromagnetic delta phase, the material as a whole remains classified as paramagnetic.
- Density: While specific figures are not given, the high aluminum content yields a lower density than other ferrous cast irons [1–5].

5.6.7 *Production, Challenges, and Critical Phenomena*

Manufacturing Chugal cast irons is complex and uncommon commercially. The main challenges are:

- Melting and Molding Problems: Aluminum's strong affinity for oxygen causes significant issues during melting and pouring, such as slag formation, oxide inclusions, and misruns ("cold shuts").
- Self-Decaying Phenomenon: A critical risk in high-aluminum irons is the possible formation of aluminum carbide (Al_4C_3). This carbide is unstable in the presence of moisture, reacting to form aluminum hydroxide ($Al(OH)_3$). The resulting volumetric expansion generates internal stresses and can lead to spontaneous disintegration and crumbling of the part. Careful control of composition and process is essential to avoid this harmful phase [1–5].

5.6.8 *Industrial Applications*

Given its extreme properties, Chugal's applications are confined to environments where oxidation resistance at very high temperatures is the primary requirement and impact loads are absent. It is used in:

- Severe oxidation environments without mechanical shock.
- Cases where behavior against hot/sulfurous gases is advantageous [1–5].

In summary, Chugal is a "superalloy" among cast irons in terms of heat resistance, but its practical use is severely limited by brittleness and manufacturing difficulties [1–5].

5.6.8.1 Duriron (High Si, Resistance to Acid Corrosion)

When the key requirement is not oxidation at 1000 °C but corrosion resistance in acidic media (at room or moderate temperatures), the reference is Duriron (~14.5% Si).

- Resistance mechanism: High Si promotes passivation via a silica-like film (iron silicates). This layer forms after surface leaching of Fe, leaving a Si-rich network that oxidizes and acts as a dense, adherent barrier against attack.
- Quantitative limits (aqueous media):
  - Sulfuric acid (H_2SO_4): practically immune even at boiling for concentrations >50%.
  - Nitric acid (HNO_3): very resistant at almost all concentrations.
  - Hydrochloric acid (HCl): limited resistance.
  - HF and H_2SO_3 (sulfurous): no useful resistance [1–5] (Fig. 5.9).

Comparison of Chugal and Duriron Cast Irons

Characteristic	Chugal	Duriron
Primary Advantage	Oxidation resistance at high temperatures	Corrosion resistance in acidic media
Alloying Element	Aluminum (20–24%)	Silicon (~14.5%)
Matrix	Ferritic	Not specified
Graphite	Free graphite	Not specified
Brittleness	High	Not specified
Machinability	Difficult	Not specified
Magnetic Properties	Paramagnetic	Not specified
Density	Lower	Not specified
Manufacturing	Complex	Not specified
Applications	High-temperature oxidation environments	Acidic media corrosion environments

Fig. 5.9 Comparison table of Chugal and Duriron cast irons, summarizing primary advantage (oxidation resistance vs. corrosion resistance), dominant alloying element (high Al vs. high Si), matrix/graphite notes, brittleness, machinability, magnetic behavior, density, manufacturing complexity, and typical application environments

5.7 Ferritic Cast Iron: Duriron (High Silicon Content 12–16%)

5.7.1 *Introduction, Composition, and Historical Context*

Duriron cast iron is a highly specialized iron alloy recognized for its exceptional resistance to corrosion in acidic media. It is classified as a ferritic cast iron with free graphite, characterized by a very high silicon content, typically 12–16%, and low carbon, 0.4–1.0%. British Standard BS 1591:1949 specifies 14.25–15.25% Si and 0.35–1.0% C. Discovery is attributed to the French metallurgist Jouve in 1908. The high silicon content is the determining factor for its unique properties. Metallurgically, silicon acts as a powerful ferrite stabilizer (alpha-former), ensuring that the alloy's microstructure is always ferritic—even during solidification—preventing austenite formation. Unlike Silal, which has lower silicon (5.5–7%), Duriron is much harder, more brittle, and practically unmachinable, destined for applications where chemical resistance is an absolute priority over mechanical properties [1–5].

5.7.2 *Microstructure and Metallurgy*

Duriron's microstructure consists of interdendritic graphite dispersed in a ferrite solid solution of silicon (silico-ferrite). During cooling, the alloy solidifies forming a eutectic of delta iron (δ) and interdendritic graphite. At room temperature, this structure remains, without the phase transformations that occur in low-silicon cast irons. Graphite may appear as fine flakes, coarse flakes, or in an acicular morphology. Occasionally, if cooling is rapid, carbides may appear in the microstructure. The alloy's most important property—its corrosion resistance—stems from forming a passive surface film of silicon oxide (SiO_2) or iron silicate. This occurs when iron atoms are leached from the silico-ferrite lattice, leaving a surface network of silicon atoms that oxidizes to form a protective, dense, adherent barrier that inhibits further corrosion [1–5] (Fig. 5.10).

Fig. 5.10 Unetched metallographic image of Duriron cast iron, showing graphite distribution within a silico-ferrite matrix typical of high-silicon corrosion-resistant irons

5.7.3 *Physical, Mechanical, and Corrosion Properties*

5.7.3.1 Mechanical and Physical Properties

Duriron's mechanical properties are dominated by extreme brittleness, a direct consequence of the silico-ferrite matrix.

- Hardness: Very hard, 450–520 HB.
- Tensile Strength: Very low, ~90–180 MPa (6–10 ton/in.^{2}).
- Toughness and Shock Resistance: Extremely low toughness. Very poor resistance to both mechanical and thermal shock, much more brittle than Silal. Izod impact strength only 4–6 ft lb (5–8 J).
- Machinability: Practically unmachinable; finishing is almost exclusively by grinding.
- Other Physical Properties:
  - Density: 7.0 g/cm^3.
  - Coefficient of Thermal Expansion: 12.2–16.3 × 10^{-6}/°C.
  - Modulus of Elasticity: 18.0 × 10^6 lb/in.^{2} (124 GPa).
  - Melting Point: For iron with 14.3% Si, ~1180 °C [1–5].

5.7.3.2 Corrosion Resistance

This is Duriron's defining characteristic:

- Mineral Acids: Extraordinary resistance to oxidizing acids such as sulfuric (H_2SO_4) and nitric (HNO_3) at practically all concentrations and temperatures, even boiling. Resistance to boiling sulfuric acid is excellent, especially above 50% concentration.
- Limitations: Limited resistance to hydrochloric acid (HCl). No useful resistance to hydrofluoric acid (HF), which dissolves the protective silica film. Also susceptible to spontaneous cracking in contact with oleum (fuming sulfuric acid).
- Organic Acids and Alkalis: Resistant to organic acids at any concentration and temperature. Behavior in alkalis is inferior to common gray iron.
- Other Corrosive Uses: Useful for handling mine waters and a wide variety of acidic wastes, effluents, and wash waters [1–5].

5.7.4 *Production and Processing*

Manufacturing Duriron parts is complex and requires strict control to avoid defects.

- Melting: Though theoretically economical, electric or rotary furnaces are preferred to control temperature and avoid carbon pickup. The practice of remelting Duriron ingots helps obtain sound castings with low gas content. Adding phosphorus (0.7–1.0%) can improve fluidity in cupola melting.

- Molding and Pouring: Parts are very prone to gas porosity (hydrogen), so completely dry raw materials and molds are vital. Solidification shrinkage is high, with a pattern contraction allowance of 1:64 (~1.6%). Part design is critical: avoid sharp corners and abrupt section changes, using generous radii to minimize cracking from internal stresses.
- Heat Treatment: To relieve internal stresses and prevent spontaneous fracture, annealing is essential: heat to 800–850 °C, hold 4–5 h, then cool very slowly in the furnace for ~30 h to room temperature.
- Welding: Very difficult but possible, requiring filler of the same composition, preheat above 700 °C, and very slow, controlled cooling to avoid cracking [1–5].

5.7.5 *Industrial Applications*

Thanks to unmatched acid resistance and relatively low cost, Duriron is used in very specific industrial niches:

- Chemical Industry: Drain piping, pumps, valves, acid concentration towers, mixing nozzles, and other fittings in chemical plants.
- Laboratories and Hospitals: Drainage systems handling corrosive chemicals.
- Other Industries: Fertilizer, explosives, paper, pigments and dyes, and electrochemical processes.
- Cathodic Protection: Widely used for anodes in impressed-current cathodic protection systems, especially in aggressive environments such as seawater or chlorinated soils [1–5].

5.7.5.1 Durichlor Variant (Duriron + Mo)

Durichlor is a commercial variant of Duriron with ~3% molybdenum (Mo). The sole purpose is to improve corrosion resistance to hydrochloric acid (HCl). Note that molybdenum does not improve the alloy's mechanical properties; Durichlor remains just as hard, brittle, and low in shock resistance as standard Duriron. Its use is therefore justified in specific applications where HCl is the main corrosive agent. There are also variants such as "Durichlor 51" that may contain chromium (~4%) to improve resistance in the presence of oxidants in HCl [1–5] (Fig. 5.11).

Duriron Cast Iron Properties

Property	Duriron	Durichlor
Silicon Content	12–16%	12–16%
Molybdenum Content	None	~3%
Hardness	450–520 HB	450–520 HB
Tensile Strength	~90–180 MPa	~90–180 MPa
Toughness	Extremely low	Extremely low
Machinability	Practically unmachinable	Practically unmachinable
Corrosion Resistance	Excellent to oxidizing acids	Improved HCl resistance

Fig. 5.11 Property comparison table for Duriron and the Durichlor variant (Duriron + Mo), contrasting Si and Mo contents, hardness, tensile strength, toughness, machinability, and corrosion-resistance emphasis (notably improved HCl resistance for Durichlor)

5.8 Austenitic Cast Iron with Free Graphite (Ni-Resist)

5.8.1 Introduction and General Classification

Ni-Resist cast irons are a family of high-alloy cast irons recognized for a microstructure with an austenitic matrix stable at room temperature. The term "Ni-Resist" is a registered trade name of International Nickel Ltd., but it has become the generic designation for these alloys. Their main characteristic is the addition of gamma-forming elements, chiefly nickel (Ni), in sufficient amounts (generally 13–36%) to suppress transformation of austenite (γ-iron) to ferrite or pearlite during cooling. This results in a ductile, tough matrix at room temperature.

These cast irons are divided into two broad families according to the morphology of free carbon in their structure:

- Lamellar-Graphite (Gray) Ni-Resist: Essentially austenitic gray irons, standardized by specifications such as ASTM A436. The graphite appears as flakes. They are identified in ISO/EN standards with designations such as L-NiCuCr 15 6 2.

- Spheroidal-Graphite (Ductile) Ni-Resist: Austenitic ductile irons, standardized by specifications such as ASTM A439. The graphite is nodular/spheroidal, which gives them much higher mechanical properties, especially ductility and toughness. They are identified with designations such as D-2 or S-NiCr 20 2.

They are used where a combination of corrosion resistance, heat resistance, dimensional stability, nonmagnetic properties, and low-temperature toughness is required [1–5].

5.8.2 Chemical Composition and Role of Alloying Elements

Ni-Resist compositions vary widely to optimize specific properties. General ranges are: 2–3% C, 1–6% Si, 0.5–2% Mn, 13.5–36% Ni, 1–6% Cr, and 0.5–7% Cu.

- **Nickel (Ni)**: The principal, essential alloying element. Stabilizes austenite, acts as a graphitizer, and improves corrosion resistance in reducing and alkaline media. Its content largely determines the coefficient of thermal expansion.
- **Chromium (Cr)**: Added to form complex carbides (M_3C and M_7C_3) that increase hardness, wear resistance, erosion resistance, and high-temperature stability.
- **Copper (Cu)**: Improves corrosion resistance in weak acids, seawater, and sulfur-bearing environments. However, copper hinders or prevents graphite nodularization, so ductile Ni-Resist grades are usually copper-free or very low in Cu (<0.5%).
- **Silicon (Si)**: Improves fluidity in casting and high-temperature oxidation resistance by promoting an adherent oxide layer. Very high contents (5–6%) give rise to the Nicrosilal variant, with excellent resistance to dry corrosion.
- **Manganese (Mn)**: Also an austenite stabilizer and used to partially replace nickel in low-cost alloys such as Nomag (~10% Ni, ~5% Mn), although these are more brittle and less corrosion-resistant.
- **Phosphorus (P)**: Must be kept very low (<0.08%) to avoid low-melting phosphide eutectics (~950 °C) that embrittle the alloy and limit its service temperature [1–5].

5.8.3 Microstructure

A Ni-Resist microstructure consists of graphite (lamellar or spheroidal) dispersed in a fully austenitic matrix. Depending on composition—especially chromium content—complex interdendritic carbides may also be present, though their fraction is usually limited to 5–10% by volume.

The austenitic matrix is stable from cryogenic temperatures up to near its melting point, so these irons do not undergo phase transformations (such as ferrite–austenite or martensitic) during heating or cooling. This gives them great dimensional stability and notable resistance to thermal cycling. Inoculation is used to refine graphite

Fig. 5.12 Metallographic image of Ni-Resist cast iron with lamellar graphite etched with Marshall reagent, showing graphite flakes dispersed in an austenitic matrix

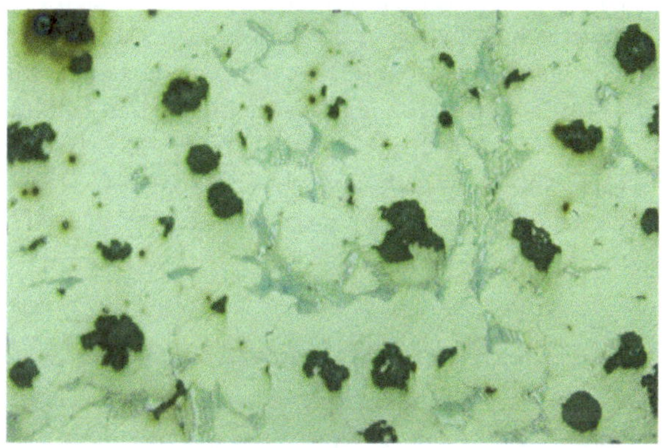

Fig. 5.13 Metallographic image of Ni-Resist cast iron with spheroidal graphite etched with Marshall reagent, showing nodular graphite particles in an austenitic matrix

morphology: in lamellar grades, Type A (random) or Type D (undercool) graphite is sought to improve heat and corrosion resistance. In ductile grades, magnesium treatment produces spheroidal graphite [1–5] (Figs. 5.12 and 5.13).

5.8.4 Mechanical Properties

- Strength and Hardness: Lamellar-graphite Ni-Resist have moderate tensile strength, 140–240 MPa, with hardness 100–250 HB. Ductile (spheroidal graphite) versions are significantly stronger, with 370–500 MPa tensile strength and similar hardness (130–270 HB). The austenitic matrix tends to work-harden, which complicates machining with light finishing cuts.
- Ductility and Toughness: A major differentiator. Lamellar Ni-Resist are brittle, with 1–3% elongation. Ductile Ni-Resist show considerable ductility, with elongations from 7% to 40%. An outstanding property is excellent low-temperature

toughness. Being austenitic, they do not exhibit a ductile-to-brittle transition, maintaining toughness even at cryogenic temperatures (to −196 °C), unlike ferritic or pearlitic cast irons.
- Wear Resistance: Good resistance to frictional wear and galling, thanks to the combination of a tough austenitic matrix and graphite lubricity. Higher-chromium types (D-2B, D-3, D-4) are recommended for severe cavitation erosion [1–5].

5.8.5 *Physical and Thermal Properties*

- Density: Slightly denser than gray irons, 7.3–7.7 g/cm^3.
- Thermal Conductivity: Lamellar-graphite versions have high thermal conductivity (~38–42 W/m K), similar to gray iron. Ductile versions, however, have much lower conductivity (~12.6 W/m K), comparable to austenitic stainless steels.
- Thermal Expansion (CTE): Very wide range. Most grades have high CTE, 18–19.3 × 10^{-6} °C^{-1}, making them susceptible to thermal shock if not managed. However, the Minvar variant (Type 5, ~35% Ni) was designed to have very low CTE, similar to carbon steel (~12 × 10^{-6} °C^{-1}) or even lower (~5 × 10^{-6} °C^{-1}), dramatically improving thermal-shock resistance [1–5].

5.8.6 *Electrical and Magnetic Properties*

- Electrical Resistivity: High, between 140 and 170 μΩ cm.
- Magnetic Properties: Most grades are nonmagnetic (paramagnetic), with relative magnetic permeability near 1.0 ($\mu_r \approx 1.02$). However, carbides can induce slight magnetism. Minvar (Type 5) and Nomag (~10% Ni, ~5% Mn) are exceptions: Minvar is ferromagnetic, while Nomag is a low-cost alternative designed specifically for nonmagnetic properties [1–5].

5.8.7 *Corrosion and Heat Resistance*

- Corrosion Resistance: Far superior to unalloyed cast irons; effective in seawater, brines, alkaline solutions, and weak acids. Performance intermediate between ductile irons and stainless steels. Useful where product contamination must be avoided (e.g., food industry).
- High Temperature: Maintain useful mechanical properties up to ~800 °C and resist oxidation and growth up to ~815–950 °C (depending on grade). In sulfurous atmospheres, limit to ~540–575 °C to avoid low-melting Ni–S eutectics.

Adding ~1% Mo markedly raises creep and rupture strength, reaching levels comparable to or exceeding those of cast stainless steels.
- Low Temperature (Cryogenics): Ductile variants retain excellent toughness at very low temperatures without ductile–brittle transition; e.g., D-2C can absorb ~20.3 J at −196 °C (Charpy V-notch).
- Manufacturing Note: The high Cu content of some variants (e.g., Type 1) is incompatible with nodularization; spheroidal graphite will not be obtained with those compositions [1–5].

5.8.8 Processing: Heat Treatment, Machinability, and Welding

- Heat Treatments: Generally used as-cast. The most common treatments are stress relief (620–675 °C) or stabilization for high-temperature service (760–870 °C). They are not hardened by conventional quenching.
- Machinability: Good to excellent, comparable to austenitic stainless steels but aided by the presence of graphite. The austenitic matrix work-hardens, so light finishing cuts should be avoided.
- Welding: Considered difficult but feasible. The preferred method is arc welding with nickel–iron electrodes (55% Ni), often with preheating to avoid cracking. Type D-2W was developed with niobium (Nb) to improve weldability [1–5].

5.8.9 Industrial Applications

Ni-Resist irons are used across a wide range of industries due to their versatility:

- Chemical and Petrochemical: Pumps, valves, compressors, and piping for brines, caustics, weak acids, and oilfield fluids.
- Marine: Seawater pumps, valves, and components exposed to saline corrosion.
- Internal Combustion Engines: Cylinder liners, valve guides, ring carriers in aluminum pistons, exhaust manifolds, and turbocharger/turbine housings.
- Low- and High-Temperature Service: Valves and components for cryogenic service, as well as furnace parts, grates, and supports for heat treatment.
- Precision and Nonmagnetic Applications: Bases for measuring instruments, glass molds (Minvar variant), and components for electrical machinery where nonmagnetism is crucial (Nomag variant).
- Food Industry: Processing equipment to avoid contamination and discoloration of products [1–5].

Ni-Resist Cast Iron Comparison

Characteristic	Lamellar-Graphite	Spheroidal-Graphite
Graphite Morphology	Flakes	Nodular/Spheroidal
Tensile Strength	140–240 MPa	370–500 MPa
Elongation	1-3%	7-40%
Thermal Conductivity	~38–42 W/m·K	~12.6 W/m·K
Standard	ASTM A436	ASTM A439

Fig. 5.14 Ni-Resist comparison table (lamellar vs. spheroidal graphite), correlating graphite morphology with tensile strength, elongation, thermal conductivity, and the associated ASTM standards (A436 and A439)

5.8.10 Variant for Cryogenic Service

There is a highly specialized variant of Ductile Ni-Resist, covered by ASTM A571, designed specifically for pressure-containing components in cryogenic service (down to −196 °C).

- Composition and Microstructure: Distinguished by a high nickel content and a significant manganese addition: 21.0–24.0% Ni and 3.75–4.5% Mn. This synergistic combination of gamma-forming elements ensures a fully austenitic, very stable matrix that shows no ductile–brittle transition, thus maintaining excellent toughness and impact resistance at extremely low temperatures.
- Applications: Used in manufacturing valves, pumps, and other fittings for the liquefied natural gas (LNG) industry and other cryogenic processes where structural reliability in the cold is critical [1–5] (Fig. 5.14).

5.9 Austenitic Cast Iron: Nicrosilal

5.9.1 Introduction, Classification, and Metallurgical Context

Nicrosilal is a high-alloy cast iron belonging to the family of austenitic cast irons with free graphite. It is a trade name, originally developed by the British Cast Iron Research Association (BCIRA), used for a specific variant of Ni-Resist cast irons characterized by a high silicon content, typically around 6%. Metallurgically, it was conceived as an austenitic version of the ferritic Silal cast iron, combining the austenite-matrix stability conferred by nickel with the exceptional oxidation resistance of silicon. In international standards it falls under austenitic cast iron specifications, such as BS 3468:1962 (Grade AUS 104) or, depending on graphite morphology, ASTM A436 (lamellar) and ASTM A439 (spheroidal). Its main purpose is service under heat and corrosion conditions, especially at temperatures above 700 °C [1–5].

5.9.2 Chemical Composition and Role of Alloying Elements

Nicrosilal's composition is designed to maximize high-temperature performance. Although variations exist, a high-performance formulation cited in the sources is 32% Ni–6% Si–5% Cr. More general ranges are: 1.6–3% C, 4.0–6.0% Si, 18–32% Ni, and 1.8–5% Cr, with other elements such as Mn, Mo, and Cu in smaller amounts.

- **Nickel (Ni)**: The fundamental element that stabilizes the austenitic matrix at room and low temperatures, conferring toughness and thermal-shock resistance.
- **Silicon (Si)**: Responsible for the excellent oxidation and growth resistance at high temperatures. Silicon promotes the formation of a very adherent, protective surface oxide layer that prevents subsurface oxidation. Contents above 5%, however, can lead to formation of a brittle silico-ferrite phase.
- **Chromium (Cr)**: Increases heat and wear resistance by promoting the formation of complex carbides in the microstructure. These same carbides are the reason Nicrosilal is more brittle than low-chromium Ni-Resist grades [1–5].

5.9.3 Microstructure

Nicrosilal's microstructure consists of free graphite dispersed in a fully austenitic matrix. For the best mechanical properties and oxidation resistance, fine graphite is sought, preferably undercooled Type D. A distinctive feature of this alloy, due to its high chromium and silicon, is the presence of complex chromium carbides in the microstructure, especially if Cr exceeds 1.8%. These carbides contribute to hardness and wear resistance but also to brittleness. If silicon exceeds 5%, an additional silico-ferrite phase may form, which is also brittle and should be avoided [1–5].

5.9.4 Mechanical and Physical Properties

- *Mechanical Properties*: Nicrosilal shows tensile strength of 170 to 278 N/mm^2 (11–18 ton/in.^{2}), elongations up to 3%, and Brinell hardness of 140 to 190 HB. Although mechanically robust for a cast iron, it is more brittle than other Ni-Resist due to its carbides.
- *Physical Properties*:
  - Density: 7.2–7.4 g/cm^3.
  - Coefficient of Thermal Expansion: High, 18×10^{-6}/°C, similar to other Ni-Resist grades, implying some sensitivity to thermal shock if temperature gradients are not managed.
  - Thermal Conductivity: Relatively low—0.063 cal/cgs at 100 °C (≈26 W/m K) and 0.070 cal/cgs at 430 °C (≈29 W/m K).
  - Electrical Resistivity: High, 160 μΩ cm (1.6 μΩ m).
  - Magnetic Permeability: Very low, between 1.1 and 1.4, making it practically nonmagnetic [1–5].

5.9.5 Heat and Corrosion Resistance, and Service Limitations

Nicrosilal's chief virtue is its high-temperature behavior. Its resistance to dry corrosion and oxidation above 700 °C is very high, comparable to refractory stainless steels of the 25Cr–20Ni type. This excellent behavior is due to the very adherent oxide layer that prevents subsurface oxidation, thus avoiding "growth" or swelling of the part. Its resistance to 1% and 5% sulfuric acid is roughly one hundred times that of common gray iron.

However, there are important limitations. Its high-temperature creep resistance is lower than that of refractory steels, which must be considered in applications under sustained load. There is also a critical warning for intermediate temperatures: if a Nicrosilal with Si > 5%, Cr < 1.8%, and Ni <17% is heated in the 400–700 °C range, it may undergo phase changes that make it magnetic, hard, and prone to cracking. Prolonged exposure between 400–650 °C can also cause slight growth and embrittlement [1–5].

5.9.6 Processing and Applications

Nicrosilal is readily weldable using filler rods of the same composition. As for machinability, the "hard" variety (with high chromium) requires carbide tooling. Its applications focus on components that must withstand severe heat and corrosion, such as molds for metal melting and various furnace parts and other equipment operating at high temperatures. Its combination of oxidation resistance, structural stability, and good castability makes it a valuable choice for complex-geometry parts in demanding thermal service [1–5] (Fig. 5.15).

Fig. 5.15 Nicrosilal cast iron properties infographic summarizing composition (Ni–Si–Cr system), microstructure (austenitic matrix with graphite and carbides), tensile strength, density, thermal expansion, thermal/electrical properties, magnetic permeability, heat and corrosion resistance, weldability, machinability, and applications

Nicrosilal Cast Iron Properties

Property	Description
Chemical Composition	1.6–3% C, 4.0–6.0% Si, 18–32% Ni, 1.8–5% Cr
Microstructure	Free graphite in austenitic matrix, chromium carbides
Tensile Strength	170 to 278 N/mm^2
Density	7.2–7.4 g/cm^3
Thermal Expansion	18 × 10^{-6} /°C
Thermal Conductivity	0.063 cal/cg·s at 100 °C
Electrical Resistivity	160 μΩ·cm
Magnetic Permeability	1.1 to 1.4
Heat Resistance	High above 700 °C
Corrosion Resistance	High, especially to sulfuric acid
Weldability	Readily weldable
Machinability	Requires carbide tooling for hard variety
Applications	Molds, furnace parts

5.10 Austenitic Cast Iron: Minvar (Low Coefficient of Thermal Expansion)

5.10.1 Introduction, Classification, and Metallurgical Context

Minvar cast iron is a highly specialized iron alloy within the family of austenitic cast irons with free graphite, specifically among the Ni-Resist group. Its most distinctive feature—and the main reason for its development—is its low coefficient of thermal expansion (CTE), a property derived from the well-known Invar effect in iron–nickel alloys with nickel around 36%.

In standard classifications, Minvar corresponds to ASTM A436 Type 5 when produced with lamellar graphite, and to ASTM A439 Types D-5 or D-5B if produced with spheroidal graphite. Unlike other Ni-Resist irons optimized for corrosion or heat resistance, Minvar is formulated specifically to offer maximum dimensional stability against temperature changes. Mechanically, it is considered the "softest" variant in the Ni-Resist family [1–5].

5.10.2 Chemical Composition and Role of Alloying Elements

Minvar's chemistry is tightly controlled to maximize the Invar effect. A typical formulation is:

- Carbon (C): ~2.4% max.
- Silicon (Si): 1.0–2.0%
- Manganese (Mn): 0.5–1.5% (other sources indicate max. 0.5%)
- Nickel (Ni): 34.0–36.0%
- Copper (Cu): Absent or minimal
- Chromium (Cr): Absent or minimal (<0.1%)

Nickel at 34–36% is fundamental, stabilizing the austenitic matrix and inducing the low-expansion effect. The deliberate absence of copper and chromium is crucial, since these elements—though beneficial in other Ni-Resist grades—interfere with and reduce nickel's low-expansion effect. Carbon and silicon are kept at levels that ensure graphite formation, avoiding white-iron solidification without compromising integrity of the austenitic matrix [1–5].

5.10.3 Microstructure and Distinctive Physical Properties

Minvar's microstructure consists of graphite (lamellar or spheroidal, depending on process) dispersed in a fully austenitic matrix. Due to the absence of chromium and other strong carbide formers, the carbide fraction is very low, contributing to lower hardness compared with other Ni-Resist grades. As with other austenitic cast irons, the structure is stable at very low temperatures, so there is no ductile–brittle transition and toughness remains good even in cryogenic environments.

A very important microstructural and physical characteristic of Minvar is that, unlike most other Ni-Resist cast irons (which are nonmagnetic), this alloy is ferromagnetic at room temperature. This property is directly related to the Fe–36%Ni composition, which lies in a phase-diagram region where austenite shows magnetic ordering at room temperature [1–5].

5.10.4 Mechanical, Physical, and Thermal Properties

5.10.4.1 Mechanical Properties

- Tensile Strength: Moderate, 138–172 N/mm^2 (20,000–25,000 lb/in.^{2}).
- Hardness: The softest of the Ni-Resist grades, 100–125 HB.
- Toughness and Impact: High impact resistance, far above gray iron. Tests have measured 150 ft lb (203 J), compared with 25–35 ft lb (34–47 J) for conventional gray iron. Maintains good toughness at cryogenic temperatures.
- Modulus of Elasticity: Low, ~10.5×10^6 lb/in.^{2} (72 GPa) [1–5].

5.10.4.2 Thermal Properties (Key)

- Coefficient of Thermal Expansion (CTE): This is its standout property. Values are exceptionally low and temperature-range dependent:
  - $3.0–4.0 \times 10^{-6}$/°C for 10–93 °C.
  - $5.0–6.0 \times 10^{-6}$/°C for 10–205 °C.

- Other sources give a more general average of ~12×10^{-6}/°C, similar to carbon steel, making it ideal for components that must maintain dimensional compatibility with steel structures [1–5].

5.10.4.3 Other Physical Properties

- Density (Specific Gravity): 7.6 g/cm^3.
- Melting Point: ~1230 °C.
- Electrical Resistivity: High, 100–170 μΩ cm [1–5].

5.10.5 Industrial Applications

Minvar's unique properties—especially dimensional stability—suit it to precision applications and environments with thermal cycling:

- Precision Components: Bases for measuring instruments, precision machine-tool components, and scientific equipment requiring high dimensional stability.
- Glass Molds: A very common application; its low, predictable CTE minimizes thermal stresses during mold heating/cooling cycles, increasing tool life and product quality.
- Structural Components: Any application needing dimensional compatibility with carbon-steel parts under thermal cycles.

Fig. 5.16 Minvar cast iron properties infographic summarizing composition limits (notably high Ni), tensile strength and hardness, impact resistance, modulus, low thermal expansion (CTE ranges), density, melting point, electrical resistivity, and magnetic behavior

Minvar Cast Iron Properties

Property	Value
Carbon (C)	~2.4% max.
Silicon (Si)	1.0–2.0%
Manganese (Mn)	0.5–1.5%
Nickel (Ni)	34.0–36.0%
Copper (Cu)	Absent or minimal
Chromium (Cr)	Absent or minimal (<0.1%)
Tensile Strength	138–172 N/mm^2
Hardness	100–125 HB
Impact Resistance	150 ft·lb
Modulus of Elasticity	~10.5 × 10^6 lb/in^2
LTE CTE (10–93 °C)	3.0 to 4.0 × 10^{-6} /°C
CTE (10–205 °C)	5.0 to 6.0 × 10^{-6} /°C
CTE (Average)	~12 × 10^{-6} /°C
Density	7.6 g/cm^3
Melting Point	~1230 °C
Electrical Resistivity	100–170 μΩ·cm
Magnetic Property	Ferromagnetic

- Cryogenic Applications: Good low-temperature toughness combined with low CTE makes it suitable for components in cryogenic systems [1–5] (Fig. 5.16).

5.11 Austenitic Cast Iron: Nomag (Nonmagnetic)

5.11.1 *Introduction, Classification, and Purpose*

Nomag cast iron is a high-alloy iron belonging to the family of austenitic cast irons with free graphite, being a specific, low-cost variant of Ni-Resist cast irons. Its name, a trademark of Ferranti Ltd., derives from its principal, defining characteristic: being nonmagnetic. Metallurgically, its distinctive trait is the partial substitution of nickel with manganese as the austenite-stabilizing element. This modification yields a fully austenitic matrix at significantly lower cost than traditional Ni-Resist grades that rely solely on high nickel content. However, this substitution entails sacrifices in other key properties—corrosion resistance and toughness—which are inferior to those of high-nickel Ni-Resist. The alloy is covered by international standards such as ISO L-NiMn13-7 and EN-GJLA-XNiMn13-7 [1–5].

5.11.2 *Chemical Composition and Role of Alloying Elements*

Nomag's composition is carefully balanced to ensure a stable austenitic matrix at the lowest possible cost. Sources provide several composition ranges, reflecting its evolution and patent specifications:

- General Composition: 2.5–3.0% C; 1.5–3.0% Si; 4.0–7.0% Mn; 11.0–12.0% Ni.
- Patent Composition (broader): 3.0–3.8% C; 2.0–3.1% Si; 6–11% Mn; 0.5–2.5% Cu; 0.1–1.0% Al.
- Simplified Composition (common reference): 10% Ni and 5% Mn [1–5].

The synergistic role of the alloying elements is:

- Nickel (Ni) and Manganese (Mn): Both are gamma formers, i.e., they stabilize austenite (γ-phase) at room temperature. Their combination depresses the martensitic transformation temperature below room temperature, ensuring the matrix remains austenitic and thus nonmagnetic.
- Manganese as Substitute: Manganese is a more economical austenite stabilizer than nickel; unlike nickel, however, it does not contribute corrosion resistance.
- Silicon (Si): Acts as a graphitizer, ensuring carbon precipitates as graphite rather than brittle carbides during solidification.
- Copper (Cu) and Aluminum (Al): In the patent formulation, copper and aluminum also contribute to austenite stabilization and graphitization [1–5].

Fig. 5.17 Metallographic image of Nomag cast iron etched with Marshall reagent, illustrating the characteristic etched microstructure of this non-magnetic austenitic cast iron family

5.11.3 Microstructure

Nomag's microstructure consists of free graphite dispersed in a fully austenitic matrix. The graphite may be lamellar (flake) or spheroidal (nodular), depending on the treatment applied to the liquid metal. The matrix, stabilized by the Ni + Mn combination, is free of ferromagnetic phases such as ferrite or martensite, which is essential for its nonmagnetic properties [1–5] (Fig. 5.17).

5.11.4 Mechanical and Physical Properties

- Tensile Strength and Hardness: Mechanical properties are lower than those of high-nickel Ni-Resist. Tensile strength lies between 139 and 216 N/mm^2 (9–14 ton/in.^{2}) for the lamellar-graphite version; one patent specifies a minimum strength >150 MPa. Brinell hardness is 140–190 HB, though a patent indicates a maximum of 270 HBS.
- Toughness and Brittleness: Sources agree it is more brittle than high-nickel Ni-Resist cast irons. Even so, its austenitic matrix gives it higher toughness than common gray irons.
- Density: While specific figures are not given, density is expected to be similar to other austenitic cast irons, in the 7.3–7.7 g/cm^3 range [1–5].

5.11.5 Magnetic and Electrical Properties (Key Characteristics)

- Magnetic Properties: Nomag's defining characteristic is its nonmagnetic behavior. Its relative magnetic permeability (μ_r) is very low, with values of 1.03–1.11, while a more recent patent specifies permeability <3.5. This makes it practically insensitive to magnetic fields, with behavior similar to copper or aluminum.
- Electrical Resistivity: High electrical resistivity, 130–150 μΩ cm (1.3–1.5 μΩ m). One patent indicates resistivity >1.2 μΩ m (120 μΩ cm). This high resistivity is advantageous in electrical applications, minimizing eddy-current losses [1–5].

5.11.6 Corrosion and Heat Resistance

Nomag's performance in aggressive environments is one of its main limitations versus other Ni-Resist grades.

- Corrosion Resistance: Poor. As noted, manganese does not confer the corrosion protection that nickel provides. Therefore, it is not recommended where corrosion resistance is important.
- Heat Resistance: Limited oxidation resistance, suitable for service temperatures up to approximately 500 °C. Above this temperature, oxidation accelerates [1–5].

5.11.7 Industrial Applications

Nomag's applications are directly linked to its nonmagnetic properties and relatively low cost. It is used mainly in:

- Components in electrical or magnetic fields: Ideal for structural parts of electrical machinery where ferromagnetism could interfere with performance, such as generator covers and switchgear cases.
- Low-cost alternative to nonferrous metals: Used as a lower-cost substitute for copper or aluminum in parts that must be nonmagnetic but also require some structural strength and stiffness [1–5] (Fig. 5.18 and Table 5.1).

Nomag Cast Iron Properties

Property	Nomag
Composition	2.5–3.8% C, 1.5–3.1% Si, 4–11% Mn, 0.5–12% Ni, 0–2.5% Cu, 0–1% Al
Microstructure	Free graphite in austenitic matrix
Tensile Strength	139–216 N/mm^2 (lamellar graphite)
HB Hardness	140–270 HB
Toughness	More brittle than high-Ni Ni-Resist
Density	7.3–7.7 g/cm^3 (estimated)
Magnetic Permeability	1.03–3.5
Electrical Resistivity	130–150 μΩ·cm
Corrosion Resistance	Poor
Heat Resistance	Up to 500 °C
Applications	Electrical components, low-cost alternative

Fig. 5.18 Nomag cast iron properties infographic summarizing composition (C–Si–Mn–Ni with possible Cu/Al), microstructure (free graphite in austenitic matrix), tensile strength (lamellar graphite variant), hardness range, relative brittleness, estimated density, magnetic permeability range, electrical resistivity, corrosion/heat resistance notes, and typical applications

Table 5.1 Summary of alloyed cast irons with free graphite

Type	Typical composition	Microstructure (RT)	Notable properties	Typical uses
Ni-Tensil (low-alloy pearlitic gray)	C 3.0–3.5; Si 1.5–2.9; Mn 0.6–0.9; Ni 0.5–3.5; Cr 0.3–1.0; Mo 0.2–1.0; Cu 0.2–1.0%	Flake graphite in fine pearlitic matrix	High UTS vs common grays; good wear; avoid free ferrite and massive carbides	High-strength parts where fine pearlite is desired (Ni-Tensil family)
Bainitic/acicular	Alloyed (e.g., Ni, Mo, Cu, Cr) for bainitic/acicular matrix	Free graphite with bainitic/acicular matrix	Better strength–toughness compromise than pearlitic; high UTS/toughness	Components seeking strength + toughness (acicular)
Self-hardening martensitic	Typical: C 2.9–3.3; Si 1.5–2.0; Mn 0.7–1.0; Cr 0.6–1.0; Ni 4.0–5.0%	Flake graphite + martensitic matrix with variable γR (self-hardens on mold cooling)	High hardness (≈400–500 HB after stated routes), brittle; temper/anneal options for machining	Wear without heavy blows; alternative to surface hardening when bulk hardening is desired
Silal (ferritic high Si)	Si 5.5–7.0%, C < 2.3, Mn 0.5–0.8 (opt. Cr ≤ 2, Mo ≤ 2)	Interdendritic graphite (Type D) in stable ferritic matrix; A1 ≈ 970 °C	Resists oxidation and growth up to 850–900 °C; 160–278 MPa, 140–190 HB; brittle cold, better >250 °C	Furnace grates/supports, nozzles, parts exposed to hot gases
Al–Si ferritic (4–6% Al)	C 2.5–3.5; Si 1.6–2.1; Al 4–6; Mn 0.1–0.4% (Mo up to 2% improves creep/oxidation)	Free graphite in ferritic matrix; possible nodular (bull's-eye) with Mg	Very resistant to oxidation (Al_2O_3); UTS 175–225 MPa (flake)/390–570 MPa (nodular); brittle; hard to machine	Hot furnace parts where oxidation rules and no severe shocks
Chugal (20–24% Al)	C ~ 2; Si ~ 1; Al 20–24%; sometimes Ce traces (modifier)	Free graphite in ferritic matrix; paramagnetic at >~20% Al	Maximum oxidation/heat resistance (1000–1100 °C); very brittle; Al_4C_3 risk if poorly processed	Extreme oxidation niches without mechanical impact; sulfurous/H_2SO_4 atmospheres favorable
Duriron (12–16% Si)	Si 12–16% (BS 14.25–15.25); C 0.4–1.0%	Interdendritic graphite in silico-ferrite; protective surface SiO_2 film	Outstanding chemical resistance (acids); very hard (450–520 HB) and brittle; UTS 90–180 MPa; requires stress-relief anneal; Durichlor variant (+~3% Mo) for HCl	Chemical drains, pumps/valves in chemical industry, cathodic-protection anodes

(continued)

Table 5.1 (continued)

Type	Typical composition	Microstructure (RT)	Notable properties	Typical uses
Ni-Resist (austenitic)	C 2–3; Si 1–6; Mn 0.5–2; Ni 13.5–36; Cr 1–6; Cu 0.5–7% (ductile: low Cu)	Austenite + graphite (flake or nodular); 5–10% carbides if high Cr	Tough at low T (no DBTT), good corrosion in seawater/alkalis; heat service up to 800–950 °C; weldable; high CTE except variants	Marine and chemical pumps/valving; engines (manifolds/turbines); cryogenics; precision (glass, etc.)
Nicrosilal (high-Si Ni-Resist)	Ni ~ 18–32; Si 4–6 (typ. 6); Cr 1.8–5; C 1.6–3%	Austenitic with Cr carbides for Cr > ~1.8; fine graphite preferable	Excellent oxidation/heat >700 °C; high CTE; more brittle than low-Cr Ni-Resist; beware 400–700 °C if (Si > 5, Cr < 1.8, Ni < 17) → hardens/embrittles	Furnace parts, melting molds, severe thermal service
Minvar (low CTE)	Ni 34–36; Si 1–2; C ≤ ~2.4; Mn 0.5–1.5; no Cu/Cr	Austenitic + graphite; very few carbides; ferromagnetic (Fe–36Ni)	Very low CTE (≈3–6 × 10^{-6}/°C at 10–205 °C); soft (100–125 HB); good toughness even cryogenic	Glass molds, precision bases, dimensional compatibility with steel
Nomag (nonmagnetic, Ni + Mn)	C 2.5–3.0; Si 1.5–3.0; Mn 4–7; Ni 11–12 (ref. 10Ni–5Mn)	Austenitic + graphite; very low μr (≈1.03–1.11); less tough/corrosion than high-Ni Ni-Resist	Nonmagnetic, high resistivity; mechanically weaker than high-Ni Ni-Resist; oxidation service ≈≤500 °C	Parts in electric/magnetic fields; economical non-magnetic substitute

References

1. ASM Handbook Committee. (1990). *ASM handbook, volume 1: properties and selection: irons, steels, and high-performance alloys*. ASM International.
2. Davis, J. R. (Ed.). (1996). *ASM specialty handbook: cast irons*. ASM International.
3. Stefanescu, D. M. (Ed.). (2017). *ASM handbook, volume 1A: cast iron science and technology*. ASM International.
4. Laird, G., Gundlach, R., & Röhrig, K. (2000). *Abrasion-resistant cast iron handbook*. American Foundry Society.
5. Davis, J. R. (Ed.). (2000). *Nickel, cobalt, and their alloys*. ASM International.

Chapter 6
Alloyed Cast Irons Without Free Graphite

Abstract Alloyed cast irons without free graphite are high-alloy white and semi-white irons in which carbon is combined in carbides rather than precipitating as graphite (ASM Handbook Committee, ASM handbook, volume 1: properties and selection: irons, steels, and high-performance alloys, ASM International, Materials Park, OH, 1990; Davis, ed., ASM specialty handbook: cast irons, ASM International, Materials Park, OH, 1996; Stefanescu, ed., ASM handbook, volume 1A: cast iron science and technology, ASM International, Materials Park, OH, 2017). By appropriately alloying with nickel, chromium, molybdenum and other carbide-forming and matrix-stabilizing elements, these irons can be designed to provide exceptional resistance to abrasive wear, impact–abrasion and, in some cases, corrosion and high-temperature attack (ASM Handbook Committee, ASM handbook, volume 1: properties and selection: irons, steels, and high-performance alloys, ASM International, Materials Park, OH, 1990; Davis, ed., ASM specialty handbook: cast irons, ASM International, Materials Park, OH, 1996; Stefanescu, ed., ASM handbook, volume 1A: cast iron science and technology, ASM International, Materials Park, OH, 2017; Laird et al., Abrasion-resistant cast iron handbook, American Foundry Society, Des Plaines, IL, 2000). Classical ledeburitic grades such as Ni-Hard cast irons, together with martensitic high-chromium white irons, cover most severe wear applications in mining, mineral processing, cement and power generation (Davis, ed., ASM specialty handbook: cast irons, ASM International, Materials Park, OH, 1996; Stefanescu, ed., ASM handbook, volume 1A: cast iron science and technology, ASM International, Materials Park, OH, 2017; Laird et al., Abrasion-resistant cast iron handbook, American Foundry Society, Des Plaines, IL, 2000). In addition, specialized ferritic alloys without graphite but with significant aluminum additions offer outstanding oxidation and fire resistance for service in oxidizing atmospheres at very high temperatures (Davis, ed., ASM specialty handbook: cast irons, ASM International, Materials Park, OH, 1996; Stefanescu, ed., ASM handbook, volume 1A: cast iron science and technology, ASM International, Materials Park, OH, 2017; Laird et al., Abrasion-resistant cast iron handbook, American Foundry Society, Des Plaines, IL, 2000; Fashu et al., Metals 13(11):1831, 2023). This chapter reviews the compositional design, microstructure, properties, processing and applications of the main families of alloyed cast irons without free

M. González Fernández de Castro et al., *Alloyed Cast Iron*, SpringerBriefs in Materials, https://doi.org/10.1007/978-3-032-27303-1_6

graphite, emphasizing the synergy between carbide type, matrix constitution and alloying strategy (ASM Handbook Committee, ASM handbook, volume 1: properties and selection: irons, steels, and high-performance alloys, ASM International, Materials Park, OH, 1990; Davis, ed., ASM specialty handbook: cast irons, ASM International, Materials Park, OH, 1996; Stefanescu, ed., ASM handbook, volume 1A: cast iron science and technology, ASM International, Materials Park, OH, 2017; Laird et al., Abrasion-resistant cast iron handbook, American Foundry Society, Des Plaines, IL, 2000; Fashu et al., Metals 13(11):1831, 2023).

Keywords Alloyed cast irons without free graphite · White irons · Ni-Hard · High-chromium white irons · Martensitic matrix · Ledeburitic structures · Aluminum-alloyed ferritic irons · Severe abrasive wear

6.1 Ledeburitic-Matrix Cast Irons: Ni-Hard

6.1.1 Definition, Classification, and General Context

Ni-Hard cast iron is a family of high-alloy white irons characterized by a microstructure without free graphite and a predominantly martensitic matrix. Its trade name, originally registered by International Nickel Ltd. (INCO), has become the generic designation for these alloys, recognized for their exceptional resistance to abrasion and wear. They are considered one of the oldest, most cost-effective, and most industrially important groups of high-alloy irons, produced for more than eight decades.

Within international standards, Ni-Hard cast irons are covered by ASTM A532 Class I and EN 12513. Their fracture is bright white due to the presence of metallic carbides instead of graphite. Although they are extremely hard, this property comes with low toughness, making them inherently brittle [1–5].

6.1.2 Chemical Composition and Role of the Alloying Elements

The composition of Ni-Hard is carefully balanced to suppress the formation of graphite and pearlite, ensuring solidification with carbides in a martensitic matrix. Typical ranges per ASTM A532 Class I are:

- Carbon (C): 2.5–3.6%
- Silicon (Si): <0.8% (up to 2.2% in Type 4)
- Nickel (Ni): 3.3–6.5%
- Chromium (Cr): 1.4–9.0%
- Molybdenum (Mo): <1.0% [1–5]

Each element plays a fundamental role:

- **Nickel (Ni)**: The principal alloying element; its function is to suppress transformation of austenite to pearlite during cooling in the mold. This allows the matrix to transform to martensite, ensuring high as-cast hardness. The nickel content must be balanced with section thickness to avoid pearlite formation or an excessive amount of retained austenite. Contents above 6.5% can over-stabilize austenite.
- **Chromium (Cr)**: Acts as a powerful carbide stabilizer, counteracting the graphitizing effect of nickel and silicon to guarantee white solidification. In low-chromium Ni-Hard types (1.4–4%), chromium forms M_3C-type carbides. In the Ni-Hard 4 variant, with higher chromium (7–11%), M_7C_3 carbides form, which are harder and more discontinuous, improving toughness and impact resistance.
- **Carbon (C)**: Controls the volume fraction of carbides in the microstructure. For maximum abrasion resistance, higher-carbon compositions (3.2–3.6%) are used, whereas for applications with repeated impact, lower contents (2.7–3.2%) are preferred.
- **Silicon (Si)**: In Ni-Hard 1 and 2, silicon is kept low (<0.8% or <1%) to avoid graphite formation. In Ni-Hard 4, a higher content (up to 2.2%) is allowed to facilitate formation of M_7C_3 chromium carbides.
- **Molybdenum (Mo)**: Added in small amounts (<1%) to increase hardenability in thick sections, helping to suppress pearlite without excessively stabilizing retained austenite.
- **Manganese (Mn)**: Kept low (<1.3%) because it is a stronger austenite stabilizer than nickel, which can lead to excessive retained austenite and therefore lower hardness. Although nickel can be substituted by manganese (3–5% Mn) to reduce costs, the resulting material is much more brittle [1–5].

6.1.3 Microstructure and Metallurgy

The microstructure of a Ni-Hard cast iron consists of hard eutectic carbides dispersed in a matrix that is ideally martensitic. During solidification, the alloy forms a eutectic of austenite and carbides. A significant amount of retained austenite usually remains in the as-cast structure. To reach a hardness of 550 HB at least 60% martensite is required, whereas with 80–90% martensite, values above 650 HB are exceeded [1–5].

Carbide types by alloy (ASTM A532, Class I)

- Ni-Hard 1, 2, and 3 (Types A, B, C): Predominantly M_3C carbides (alloyed cementite) forming a continuous network (thin plates). Typical hardness ≈840–1100 HV: very hard but brittle; the continuity of the eutectic facilitates catastrophic fractures.
- Ni-Hard 4 (Type D, ~7–11% Cr): Predominantly M_7C_3 carbides, harder (≈1200–1600 HV) and with a more discontinuous/isolated morphology. This discontinuity interrupts crack propagation paths, providing appreciably higher toughness and impact resistance than Types A–C [1–5].

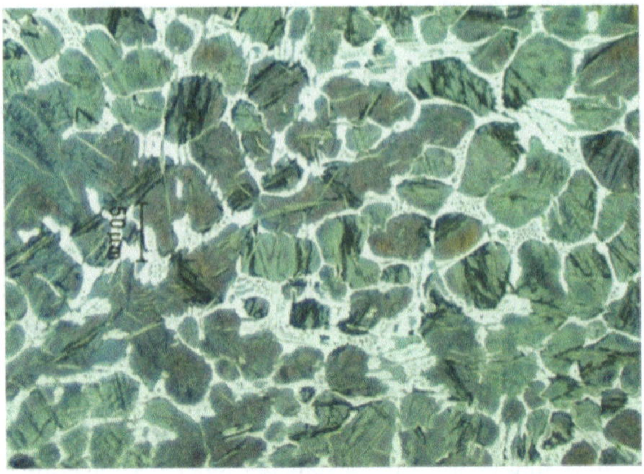

Fig. 6.1 Metallographic image of Ni-Hard white cast iron etched with Nital 3%, showing a ledeburitic structure composed of eutectic carbides dispersed in a predominantly martensitic matrix with retained austenite

Retained austenite can transform to martensite by mechanical deformation in service, contributing to surface work-hardening of the part [1–5] (Fig. 6.1).

6.1.4 Mechanical Properties

- Hardness: Its standout property. Ni-Hard reaches hardness in the range 500 to 780 DPN (Vickers), equivalent to 52–63 HRC. The Ni-Hard 4 variant can reach from 600 HB as-cast to more than 700 HB with heat treatment.
- Abrasion Resistance: Outstanding resistance to abrasive wear, making them very cost-effective. Performance is excellent under low- and medium-stress abrasion. The resistance is due to the combination of a hard matrix and a high volume of carbides. The M_7C_3 carbides of Ni-Hard 4 are harder than silica (quartz), explaining its superiority in certain environments.
- Toughness and Impact Resistance: Toughness is low, and they are considered brittle materials. Ni-Hard 4 exhibits the highest toughness in the family thanks to its discontinuous M_7C_3 carbides.
- Corrosion Resistance: Moderate. Ni-Hard 4, with higher chromium, offers superior resistance and is useful in moderately corrosive slurries, though these are not considered severe-corrosion-resistant alloys. Corrosion may be accelerated due to galvanic coupling between carbides (passive) and the matrix (active) [1–5].

6.1.5 Processing: Production, Heat Treatment, and Finishing

- Production and Casting: Although they can be melted in cupolas, induction electric furnaces are preferred for tighter composition control and to avoid contamination. Ni-Hard 4 stands out for excellent fluidity and moldability, allowing complex parts. They require a generous feeding system (risers) to compensate for 4–5% volumetric shrinkage and avoid defects.
- Heat Treatment:

  - Stress Relief/Tempering: An essential low-temperature treatment, between 205 and 275 °C (other sources cite 205–260 °C or 230–290 °C). This tempers martensite, relieves stresses, and improves strength and toughness by 50–80% without reducing hardness. It is crucial to avoid thermal shock during heating and cooling.
  - Cryogenic Treatment (Cooling): Can be applied to reduce retained austenite and increase hardness. Cooling to −70 to −180 °C can raise hardness by up to 100 HB units.
  - High-Temperature Treatment: Formerly, an austenitizing treatment at 750–790 °C was applied to Ni-Hard 4 to improve hardness if as-cast hardness was insufficient; nowadays, cryogenic treatment is more common.

- Machinability and Welding:

  - Ni-Hard cast irons are not machinable with conventional tools; finishing is limited to grinding. For operations such as drilling, mild-steel inserts can be cast into the part.
  - Repair welding is not possible without cracking. Surface hardfacing is a common technique to reclaim parts, although the deposited layer shows numerous surface cracks that usually do not penetrate the base material [1–5].

6.1.6 Industrial Applications

The choice of Ni-Hard type depends on the required balance between abrasion resistance and toughness.

- Ni-Hard 1: Used for maximum abrasion with low impact. Applications include mill liners, grinding balls, slurry pumps, roll heads, mixer augers, coke-crusher segments, classifier shoes, and brick molds.
- Ni-Hard 2 and 3: For conditions with moderate to repeated impact. Used in crusher components such as jaws and hammers.
- Ni-Hard 4: Reserved for the most severe applications combining abrasion, impact, and moderate corrosion. Typical examples are volutes and impellers in pumps handling highly abrasive slurries, segments and tires of coal pulverizer tables, and augers for asphalt pavers [1–5].

6.1.7 Main Variants (Ni-Hard 1, 2, 3, and 4)

The Ni-Hard family comprises several standardized grades (ASTM A532 Class I) designed for different balances of abrasion and impact:

- Ni-Hard 1 and 2 (Types A and B): The most common grades. Ni-Hard 1, with higher carbon (3.2–3.6%), targets maximum abrasion resistance with low impact. Ni-Hard 2, with lower carbon (2.7–3.2%), offers higher toughness for applications with moderate impact. Both feature M_3C carbides.

- Ni-Hard 4 (Type D): A significant evolution. It contains higher chromium (7–11% Cr) and nickel (4.5–7.0% Ni). This promotes formation of M_7C_3 carbides, which are harder and, crucially, of discontinuous morphology. This microstructure provides much higher toughness and impact resistance than Ni-Hard 1 and 2, as well as better corrosion resistance. It is used in the most severe applications, such as slurry pumps.

6.1.8 Low-Cost Alternative: Manganese Martensitic Cast Iron

To reduce cost, nickel can be replaced by manganese, yielding a martensitic white cast iron alloyed with 3–5% manganese. Although high hardness is achieved (up to 550 HB after air quench), these alloys are considerably more brittle than standard Ni-Hard, and their use is very limited (Fig. 6.2).

Ni-Hard Cast Iron Variants Comparison

Characteristic	Ni-Hard 1 & 2	Ni-Hard 4	Manganese Martensitic
Carbide Type	M_3C	M_7C_3	Unknown
Carbon Content	2.7–3.6%	Not specified	Not specified
Chromium Content	1.4–4%	7–11%	Not specified
Nickel Content	3.3–6.5%	4.5–7%	Replaced by Manganese
Hardness	840–1100 HV	1200–1600 HV	Up to 550 HB
Toughness	Lower	Higher	More Brittle
Corrosion Resistance	Moderate	Superior	Not specified
Abrasion Resistance	Maximum	High	Not specified
Impact Resistance	Low to Moderate	High	Not specified
Cost	Standard	Standard	Lower

Fig. 6.2 "Ni-Hard cast iron variants comparison" chart summarizing the differences between Ni-Hard 1/2, Ni-Hard 4, and manganese martensitic white iron, comparing carbide type, C/Cr/Ni contents, hardness, toughness, corrosion and abrasion resistance, impact resistance, and relative cost

6.2 Martensitic Cast Iron Without Free Graphite: High Chromium (Hi-Cr 11–28% Cr)

6.2.1 Introduction, Classification, and General Context

High-chromium white cast irons are a group of high-alloy irons designed to offer a superior combination of abrasion resistance, toughness, and corrosion resistance. They are characterized by a microstructure without free graphite and a chromium content ranging from 11% to 30%. These alloys are widely used in industrial applications where severe wear is the main failure mode, outperforming Ni-Hard irons in conditions involving impact and corrosion.

They are classified mainly under ASTM A532 into two principal categories:

- Class II (Chromium–Molybdenum, 11–23% Cr): Often called medium- to high-chromium; these alloys are alloyed with molybdenum to ensure formation of a martensitic matrix in thick sections. They are generally considered the hardest of all white-iron grades.
- Class III (High Chromium, 23–30% Cr): General-purpose alloys selected when, in addition to abrasion, a significant level of corrosion resistance is required.

The European equivalent standard is EN 12513 [1–5].

6.2.2 Chemical Composition and Role of Alloying Elements

The chemical composition of these irons is carefully adjusted to achieve the desired balance among hardness, toughness, and corrosion resistance. Typical ranges are:

- Carbon (C): 1.6–3.6%.
- Chromium (Cr): 11–30%.
- Molybdenum (Mo), Nickel (Ni), Copper (Cu): Added to increase hardenability, i.e., the ability to form a martensitic matrix instead of pearlite during air cooling [1–5].

The role of each element is crucial:

- **Chromium (Cr)**: The fundamental alloying element. At contents above 10–12%, it ensures that the primary eutectic carbide is of the M_7C_3 (or K_2) type rather than M_3C. It also prevents graphitization and, when dissolved in the matrix, imparts notable corrosion and high-temperature oxidation resistance.
- **Molybdenum (Mo)**: A very effective pearlite suppressor, increasing hardenability in thick sections without excessively stabilizing retained austenite. It also partitions between carbides and matrix, contributing to hardness.

- **Nickel (Ni) and Copper (Cu)**: Like molybdenum, used to improve hardenability and ensure a fully martensitic or austenitic matrix, suppressing pearlite.
- **Carbon (C)**: Directly controls the volume fraction of carbides. Higher contents increase hardness and abrasion resistance but decrease toughness and fracture resistance.
- **Silicon (Si)**: Must be carefully controlled. Although it aids fluidity, in excess it can promote pearlite formation and should be kept low for good heat-treatment response [1–5].

6.2.3 Microstructure and Metallurgy

The microstructure of these irons is that of an in-situ metallurgical composite, consisting of extremely hard primary and/or eutectic carbides dispersed in a tough metallic matrix.

- Carbides: The predominant carbide is M_7C_3 (or K_2), a chromium–iron carbide with a hardness of 1300–1800 HV. Unlike the M_3C carbides in Ni-Hard—which form a continuous, brittle network—M_7C_3 carbides have a rod- or plate-like, discontinuous morphology. This discontinuity is key to the superior toughness of high-chromium irons, as it interrupts crack-propagation paths. In hypereutectic alloys, large primary carbides may appear, increasing abrasion resistance but also brittleness.
- Matrix:
  - As-cast: The matrix is usually austenitic or a mixture of austenite and martensite, due to high alloying depressing transformation temperatures.
  - After heat treatment: The matrix transforms to martensite, with a variable amount of retained austenite (10–30%). The hardness of chromium martensite is 770–800 HV.
- Solidification: Hypoeutectic alloys solidify with primary austenite dendrites, followed by an austenite + M_7C_3 eutectic [1–5] (Fig. 6.3).

6.2.4 Mechanical Properties

- Hardness: Their principal property. In the as-cast state, hardness ranges from 450 to 550 HB. After hardening heat treatment, they can reach 600 HB to 850 DPN/HV (up to 67 HRC). They are considered the hardest of all white cast irons.
- Abrasion Resistance: Performance is excellent, surpassing Ni-Hard in many applications due to the higher hardness of M_7C_3 carbides compared with M_3C, especially against hard abrasives such as quartz.

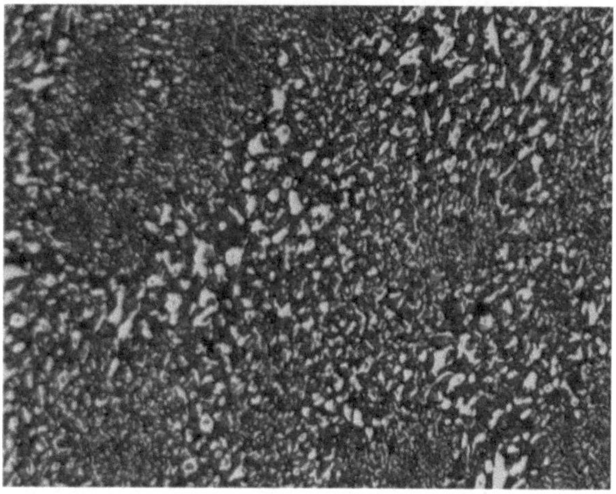

Fig. 6.3 High-chromium white cast iron etched with Nital 3%, illustrating discontinuous chromium carbides (M_7C_3) within a high-alloy matrix (typically martensitic after heat treatment)

- Toughness: Although brittle materials, their toughness is far superior to that of Ni-Hard irons. This improvement is due to the discontinuous morphology of M_7C_3 carbides, which do not provide an easy path for crack propagation. Toughness decreases as the carbide volume fraction (i.e., carbon content) increases.
- Tensile Strength: Not their main property, but higher than that of other white irons. Typical values range from 415 to 830 MPa, depending on composition and heat treatment [1–5].

6.2.5 *Corrosion and Heat Resistance*

- Corrosion Resistance: The high chromium content in the matrix (>12%) confers good to excellent corrosion resistance, similar to stainless steels. They are particularly effective in oxidizing media such as nitric and sulfurous acids [4001, from earlier table]. This makes them ideal for slurry pumps, mine waters, and brick dies [4001, from earlier table, 2450].
- Heat Resistance: Highly resistant to heat and oxidation, withstanding continuous temperatures of 800 to 950 °C. The absence of free graphite gives them great resistance to growth, and the stability of carbides at high temperature maintains their hot hardness and wear resistance [1–5].

6.2.6 *Processing and Heat Treatments*

Producing these alloys requires very strict control.

6.2.6.1 Melting and Pouring

- Prefer electric furnaces (arc or induction) for precise composition control.
- Avoid cupolas due to high carbon pickup.
- Heating rate: slow (≤30 °C/h for complex geometries) to minimize thermal shock.
- Hardening heat treatment ("destabilization quench") [1–5].

6.2.6.2 Austenitizing

- Promotes precipitation of fine secondary carbides, reducing C in austenite and destabilizing it for transformation to martensite.
- Class II (12–20% Cr): 950–1010 °C.
- Class III (23–28% Cr): 1010–1090 °C.
- Time at temperature: ≥4 h (ensures destabilization) [1–5].

6.2.6.3 Air Quench

- Cool in air (preferably with forced ventilation).
- Avoid water/oil due to high risk of cracking [1–5].

6.2.6.4 Tempering

- 200–230 °C, 2–4 h (or 200–250 °C as operating range).
- Relieves stresses, tempers martensite, and can raise toughness ~20% without significant loss of hardness.

6.2.6.5 Annealing for Machinability

- Unlike Ni-Hard, they can be annealed for machining.
- Subcritical: 690–705 °C.
- Full: 955–1010 °C.
- Goal: soften the matrix to ~400–450 HB, enabling machining operations.

6.2.7 Microconstituent Hardness (Reference)

- M_7C_3 carbides: ≈1500–1800 HV (very hard).
- Alloyed (Cr) martensite: ≈770–800 HV.

- Alloyed (Cr) austenite: ≈350–400 HV (may vary with stabilization/retention) [1–5].

6.2.8 *Industrial Applications*

Thanks to their unique combination of properties, they are used in some of the most severe wear applications:

- Mining and Mineral Processing: Mill liners, grinding balls, impact plates, and crusher components for hard rock.
- Slurry Pumps: Impellers and casings for pumping abrasive and corrosive slurries in mining, dredging, and the oil-sand industry.
- Cement and Pulverizing Industry: Rollers and tables in coal and clinker mills where high-temperature abrasion resistance is required.
- Shot-blasting Equipment: Impellers, blades, and liners.
- Recycling Industry: Impact bars and hammers in shredders for scrap, concrete, or glass.
- Other Applications: Brick-making dies, equipment for the paper industry (refiner plates), and components for the oil-sand industry [1–5] (Fig. 6.4).

Comparison of High-Chromium Cast Iron Classes

Characteristic	Class II (Cr-Mo)	Class III (High Cr)
Chromium Content	11-23%	23-30%
Molybdenum	Added	Not always added
Hardness	Hardest	High
Corrosion Resistance	Good	Significant
General Use	Thick sections	General purpose

Fig. 6.4 "Comparison of high-chromium cast iron classes" chart contrasting Class II (Cr–Mo, 11–23% Cr) and Class III (high Cr, 23–30% Cr), highlighting molybdenum addition, relative hardness, corrosion resistance, and intended general use

6.3 Ferritic Cast Iron Without Free Graphite: High Chromium (>28% Cr)

6.3.1 Definition and General Classification

High-chromium ferritic white cast irons are a group of high-alloy irons characterized by a microstructure without free graphite and a chromium content above 28%. They belong to a category of materials designed to offer an exceptional combination of resistance to severe abrasive wear, corrosion, and high temperatures.

Within the classifications of high-chromium cast irons for high-temperature service, these alloys are categorized as ferritic, specifically those with chromium contents between 30% and 34%. Their main advantage is their ability to operate under extreme conditions where other cast irons would fail, such as in pumps for abrasive slurries or in furnace components working at 800–900 °C [1–5].

6.3.2 Chemical Composition

The composition of these cast irons is carefully balanced to suppress the formation of graphite and obtain a stable ferritic matrix with chromium carbides.

- **Chromium (Cr)**: 28–35%. It is the fundamental element. Such a high content completely inhibits graphitization during solidification and, when dissolved in the matrix, confers corrosion resistance similar to that of a stainless steel.
- **Carbon (C)**: The content varies according to the application, with two main ranges:
  - Moderate carbon content (2.0–2.6%): Used for extreme abrasive-wear applications.
  - Low carbon content (approx. 1.2%): This hypoeutectic composition reduces the carbide fraction, which maximizes the amount of chromium dissolved in the matrix and therefore optimizes corrosion resistance.
- **Other Alloying Elements**: Small amounts of other elements may be added to refine the microstructure and improve specific properties:
  - Molybdenum (Mo): Up to 1–2%, to improve hardenability.
  - Nickel (Ni) or Copper (Cu): ≤1.5%, to stabilize austenite during cooling and refine the structure.
  - Silicon (Si) and Manganese (Mn): ~0.5–1.5%, for casting control and deoxidation [1–5].

6.3.3 Microstructure

The microstructure of these alloys is that of a natural metallurgical composite, consisting of extremely hard carbides dispersed in a metallic matrix resistant to corrosion and heat.

- Metallic Matrix:
  - The matrix is predominantly ferritic, especially in the annealed state or in alloys with low carbon and very high chromium.
  - Thanks to the strong ferritizing capacity of chromium, the alloy can solidify forming primary delta ferrite, similar to ferritic stainless steels.
  - In the as-cast state, the matrix may be a mixture of ferrite, austenite, and martensite, depending on the exact composition and cooling rate. The purely ferritic state is reliably achieved by an annealing treatment.
- Carbides:
  - With no graphite present, all the carbon combines to form chromium carbides. The predominant carbide type is M_7C_3, which is significantly harder than the cementite (M_3C) present in lower-alloy cast irons. In some compositions carbides of the $M_{23}C_6$ type may also appear.
  - These carbides have a fibrous or lamellar morphology and are distributed in the interdendritic network, reinforcing the matrix [1–5].

6.3.4 Properties

6.3.4.1 Mechanical Properties

- Hardness: Hardness is very high. In the as-cast state it ranges between 450–600 HB (46–55 HRC). After a hardening treatment it can reach 650–750 HB (58–63 HRC). However, by annealing they can be softened down to 350–400 HB to allow machining operations.
- Toughness: It is an inherently brittle alloy with low impact toughness, due to the high fraction of hard carbides. Its elongation is practically nil.
- Tensile and Compressive Strength: Tensile strength is modest (generally below 700 MPa), but its compressive strength is exceptional, reaching 2000–3000 MPa [1–5].

6.3.4.2 Wear, Corrosion, and High-Temperature Resistance

- Wear Resistance: Extremely resistant to severe abrasive wear, even dry or with hard particles. This property is due to the presence of M_7C_3 chromium carbides, which have a hardness of 1500–1800 HV.

- Corrosion Resistance: Its corrosion resistance is very high, especially in acidic and oxidizing media. The ferritic matrix, rich in chromium, forms a passive chromium-oxide layer that protects the material in a manner similar to a stainless steel.
- High-Temperature Resistance: They are very stable at high temperatures. They can operate continuously at 800–900 °C and in some applications up to 1040 °C (1900 °F), thanks to the stability of their ferritic matrix and the protective chromium-oxide film [1–5].

6.3.5 Processing and Heat Treatments

- *Manufacture*: Produced by casting in electric or induction furnaces for precise composition control. Mold design requires generous feeding systems to compensate for shrinkage and avoid defects.
- *Machinability*: They are not machinable in the hardened state; finishing is performed by roughing or grinding. However, they can be made machinable by an annealing treatment that softens the matrix to a hardness of 350–400 HB.
- *Heat Treatments*:
  - Annealing: Used to soften the part and facilitate machining. This treatment is what guarantees obtaining the ferritic matrix.
  - Hardening ("destabilization quench"): If maximum hardness and wear resistance are sought, an austenitizing treatment (950–1100 °C) is carried out followed by air cooling to transform the matrix to martensite. A subsequent low-temperature temper (450–550 °C) can relieve stresses and improve wear resistance [1–5].

6.3.6 Variants for High-Temperature Service

High-chromium cast irons are inherently heat-resistant, but specific variants can be designed to optimize performance at elevated temperatures (up to 1040 °C), classified according to their matrix:

- Martensitic (12–28% Cr): Offer a good balance between wear resistance and hot performance.
- Ferritic (30–34% Cr): As described in this section, they maximize structural stability and oxidation resistance at the highest temperatures.
- Austenitic (15–30% Cr and 10–15% Ni): The addition of nickel stabilizes an austenitic matrix, which improves toughness and creep resistance at high temperatures.

These alloys are used in furnace components, recuperator tubes, glass molds, and valve seats.

6.3.7 Industrial Applications

Thanks to their unique combination of properties, these cast irons are used under the most severe conditions in industry:

- Mining and Mineral Handling: Components for abrasive-slurry pumps, dredging equipment, hoppers, crushers, vertical-mill liners, and parts exposed to hard and corrosive minerals.
- High-Temperature Service: Furnace components, recuperator tubes, breaker bars and trays in sintering furnaces, grates, burner nozzles, and bottle-glass molds.
- Pulverizing and Cement Industry: Crusher hammers and mill liners processing very hard, abrasive materials.
- Oil and Drilling Industry: Drilling tools and components exposed to sand and corrosive fluids [1–5] (Fig. 6.5).

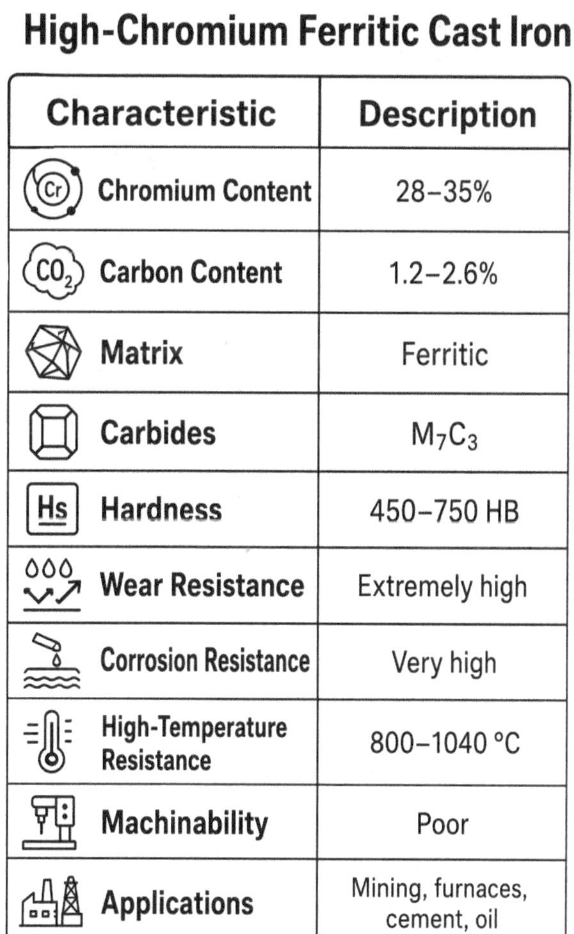

Fig. 6.5 "High-chromium ferritic cast iron" property summary chart (Cr 28–35%), listing matrix type (ferritic), dominant carbides (M_7C_3), typical hardness range, wear and corrosion resistance level, high-temperature capability, machinability, and application areas

6.4 Ferritic Cast Iron Without Free Graphite: Alloyed with Aluminum (≈7% Al)

6.4.1 Introduction, Classification, and Metallurgical Context

This alloy is a specialized high-alloy cast iron characterized by a microstructure without free graphite and a predominantly ferritic matrix, thanks to an aluminum content of approximately 7%. Unlike gray or ductile cast irons, in this alloy the carbon is in combined form, which gives it a white fracture and properties similar to those of a white cast iron.

Its main purpose is to offer excellent heat and oxidation resistance at high temperatures, being able to operate continuously up to 950 °C. It is often considered a transition material that could also be classified as "carbide–ferrite steels." Although its properties are attractive, its commercial production is uncommon, since they are difficult to manufacture due to the high reactivity of aluminum, which generates melting and molding problems and defects such as refractory slags (Al_2O_3), oxide inclusions, and incomplete filling [1–5].

6.4.2 Chemical Composition and Role of Alloying Elements

The typical composition of these cast irons is carefully adjusted to achieve the desired microstructure, with ranges such as:

- Carbon (C): 2.5–3.0%.
- Silicon (Si): ≤1%, usually around 0.3%.
- Manganese (Mn): ~0.5%.
- Aluminum (Al): ~7% (in a range of 6 to 8%) [1–5].

The role of each element is crucial:

- **Aluminum (Al)**: The key alloying element. At contents above 6.5%, it suppresses the formation of graphite. Between 7% and 18% it acts as a strong cementite stabilizer, ensuring that the carbon remains combined. In addition, it is a powerful ferrite stabilizer (alpha-forming element) and the main responsible for the exceptional high-temperature oxidation resistance, thanks to the formation of a protective and adherent aluminum-oxide (Al_2O_3) layer.
- **Carbon (C)**: Has a critical influence on mechanical properties and machinability. To obtain maximum resistance to dry corrosion, it is recommended to keep carbon low, around 1.2%. When carbon is below 1.5%, these alloys are easily machinable; however, for applications requiring higher wear resistance, carbon can be increased up to 2.5%, although this significantly hinders machining and improves castability [1–5].

Fig. 6.6 Cast iron with ~7% Al without graphite, revealing a ferritic matrix with κ-type carbides (Fe_3AlC_x). Specimen etched by 5 s immersion at room temperature in a nitric acid–hydrofluoric acid–glycerin reagent

6.4.3 *Microstructure*

The microstructure of this cast iron consists of a completely ferritic matrix containing numerous complex carbides of the Fe_3AlC_x type (known as the kappa phase) dispersed within it. The absence of free graphite is a defining characteristic.

One of its most important metallurgical properties is the absence of the allotropic phase transformation from ferrite to austenite ($\alpha \leftrightarrow \gamma$) during heating and cooling cycles. The ferritic matrix remains stable throughout its service temperature range. This stability confers one of its most notable advantages: great resistance to thermal shock, since the volume changes and internal stresses associated with phase transformations that occur in other ferrous alloys are avoided [1–5] (Fig. 6.6).

6.4.4 *Mechanical Properties*

- Strength and Hardness: The tensile strength (Rm) of these alloys lies in a range of 100–180 MPa. Their hardness is high, reaching a range of 280–320 HB.
- Toughness and Brittleness: These are brittle materials, with practically zero ductility (elongation ~0%) and low resistance to impact and shock. Their behavior is typical of a white cast iron, with brittle fracture without appreciable plastic deformation.
- Machinability: Machinability is a complex point. They are generally considered difficult to machine. However, the 7% Al variant is described as "quite well machinable" and specified as "easily machinable" if the carbon content is below 1.5%. In contrast, cast irons with more than 8% Al are very difficult to machine and often require grinding [1–5].

6.4.5 *Physical Properties*

- Density: The addition of aluminum significantly reduces density. With 7% Al, density is approximately ~6.6 g/cm^3, making them 10–15% lighter than a conventional cast iron (~7.2 g/cm^3).
- Melting Point: The solidification range is slightly elevated. The liquidus temperature is around 1250 °C and the solidus around 1170 °C.
- Coefficient of Thermal Expansion: Estimated on the order of 12–13 × 10^{-6} K^{-1}, comparable to that of a ferritic steel. The absence of graphite and of phase transformations confers homogeneous and predictable expansion.
- Thermal Conductivity: Moderate, lower than that of gray cast iron (which benefits from graphite) but comparable to that of a carbon steel, estimated at 20–30 W/m K [1–5].

6.4.5.1 Heat and Oxidation Resistance

This is the most outstanding property of these alloys. They offer excellent fire and oxidation resistance, being able to operate in continuous service at temperatures up to 950 °C. Protection is due to the formation of a dense, adherent surface layer of aluminum oxide (Al_2O_3) that is highly resistant to oxygen penetration and effectively protects the underlying metal [1–5].

6.4.6 *Electrical and Magnetic Properties*

- Electrical: They behave like a moderately alloyed ferrous metal, with an electrical resistivity estimated at 1.0–1.5 × 10^{-6} Ω m. They are not notable for conductivity.
- Magnetic: They are ferromagnetic at room temperature due to their ferritic matrix. The material is magnetically "soft," with low coercivity, meaning it magnetizes and demagnetizes easily, without retaining strong magnetism [1–5].

6.4.7 Production, Applications, and Challenges

Manufacturing these cast irons presents significant challenges due to the high reactivity of aluminum with oxygen. Induction furnaces with protective fluxes are usually employed to minimize oxide formation. They require generous feeding systems to compensate for lower fluidity and higher shrinkage.

Due to their property profile, they are used mainly in applications of moderate hot-oxidation service, such as:

- Furnace and boiler components (grates, muffles, supports).
- Heat exchangers and radiant tubes.
- Equipment for the chemical and petrochemical industry, especially in sulfurous atmospheres.
- Potentially in automotive components such as exhaust manifolds, although their brittleness is a limitation.

They are covered by standards for materials resistant to heat and corrosion, such as UNE EN 10295 [1–5].

6.4.8 High-Temperature Service Variants

High-chromium cast irons are inherently heat resistant, but specific variants can be designed to optimize performance at elevated temperatures (up to 1040 °C), classified by their matrix:

- Martensitic (12–28% Cr): Offer a good balance between wear resistance and hot performance.
- Ferritic (30–34% Cr): As described in this section, they maximize structural stability and oxidation resistance at the highest temperatures.
- Austenitic (15–30% Cr and 10–15% Ni): The addition of nickel stabilizes an austenitic matrix, which improves toughness and creep resistance at high temperatures.

These alloys are used in furnace components, recuperator tubes, glass molds, and valve seats [1–5] (Fig. 6.7).

Ferritic Cast Iron Alloyed with Aluminum (~7% Al)

Property	Description
Microstructure	Ferritic matrix with complex carbides
Carbon Content	2.5–3.0%
Aluminum Content	~7%
Tensile Strength	100 to 180 MPa
Hardness	280 to 320 HB
Ductility	~0%
Density	~6.6 g/cm^3
Maximum Service Temperature	Up to 950 °C
Machinability	Difficult, but machinable below 1.5% carbon
Magnetic Properties	Ferromagnetic, magnetically soft
Applications	Furnace components, heat exchangers

Fig. 6.7 "Ferritic cast iron alloyed with aluminum (~7% Al)" property summary chart, compiling microstructure (ferrite + complex carbides, no graphite), composition ranges, tensile strength, hardness, ductility, density, maximum service temperature, machinability conditions, magnetic behavior, and typical applications

Table 6.1 Summary of alloyed cast irons, including typical composition, room-temperature microstructure, notable properties, and typical applications

Type	Typical composition	Microstructure (RT)	Notable properties	Typical uses
Ni-Hard (ledeburitic)	C 2.5–3.6; Si < 0.8 (up to 2.2 in Type 4); Ni 3.3–6.5; Cr 1.4–9; Mo < 1%	Eutectic carbides (M_3C or M_7C_3 in Type 4) in martensitic matrix (variable γR)	Very high hardness (500–780 DPN; Ni-Hard 4 > 700 HB HT); outstanding wear; brittle; temper 205–275 °C; cryogenic useful	Mills, balls, slurry pumps, crushers, rollers, augers, etc.
Hi-Cr 11–28% Cr (martensitic)	Cr 11–30; C 1.6–3.6; +Mo/Ni/Cu for hardenability	Discontinuous M_7C_3 + martensitic matrix (after HT; γR 10–30%)	450–550 HB as-cast → 600 HB–~67 HRC after HT; better toughness than Ni-Hard; good corrosion (matrix Cr > 12%) and heat 800–950 °C	Slurry pumping, mining, brick dies, pulverizers, etc.
>28% Cr (ferritic)	Cr 28–35; C 1.2–2.6; low + Mo/Ni/Cu	Ferritic (δ/α) matrix with Cr carbides (M_7C_3/$M_{23}C_6$); no graphite	450–600 HB as-cast (650–750 HB hardened; 350–400 HB annealed); very high oxidation/corrosion; brittle	Abrasive/corrosive slurries, furnace parts 800–900 °C
≈7% Al (ferritic Al, white)	Al ~ 7; C 2.5–3.0; Si ≤ 1; Mn ~ 0.5%	Ferrite + Fe_3AlC_x (κ phase) carbides; no graphite; very stable (no α ↔ γ)	Excellent oxidation (≈950 °C); 280–320 HB; UTS 100–180 MPa; brittle; density ~ 6.6 g/cm^3; machinable if C < 1.5%	Slurry pumps, furnaces, pulverizing/cement, drilling (abrasion + heat)

6.5 Summary Table

Quick reading note: free graphite = better damping and machinability; without graphite = white fracture, maximum hardness/wear or maximum oxidation depending on the carbide system/element (Cr or Al). The matrix (pearlitic/bainitic/martensitic/ferritic/austenitic) dictates the strength–toughness balance and the temperature behavior (Table 6.1).

Cast irons with free graphite
White cast irons (without free graphite)

References

1. ASM Handbook Committee. (1990). *ASM handbook, volume 1: Properties and selection: Irons, steels, and high-performance alloys*. ASM International.
2. Davis, J. R. (Ed.). (1996). *ASM specialty handbook: Cast irons*. ASM International.
3. Stefanescu, D. M. (Ed.). (2017). *ASM handbook, volume 1A: Cast iron science and technology*. ASM International.
4. Laird, G., Gundlach, R., & Röhrig, K. (2000). *Abrasion-resistant cast iron handbook*. American Foundry Society.
5. Fashu, S., Xie, Y., Wang, J., & Xiao, W. (2023). Development and performance of high chromium white cast irons (HCWCIs) for wear/corrosive environments: A critical review. *Metals, 13*(11), 1831. https://doi.org/10.3390/met13111831

Chapter 7
Summary Table of Alloyed Cast Irons

Abstract The wide variety of alloyed cast irons can be rationalized by considering the presence or absence of free graphite, the type of metallic matrix and the total level of alloying elements (ASM Handbook Committee, ASM handbook, volume 1: properties and selection: irons, steels, and high-performance alloys, ASM International, Materials Park, OH, 1990; Davis, ed., ASM specialty handbook: cast irons, ASM International, Materials Park, OH, 1996; Stefanescu, ed., ASM handbook, volume 1A: cast iron science and technology, ASM International, Materials Park, OH, 2017). This chapter presents a concise summary table that classifies alloyed cast irons with free graphite and white irons without free graphite according to their typical composition ranges, room-temperature microstructures, key properties and main application fields (ASM Handbook Committee, ASM handbook, volume 1: properties and selection: irons, steels, and high-performance alloys, ASM International, Materials Park, OH, 1990; Davis, ed., ASM specialty handbook: cast irons, ASM International, Materials Park, OH, 1996; Stefanescu, ed., ASM handbook, volume 1A: cast iron science and technology, ASM International, Materials Park, OH, 2017; Laird et al., Abrasion-resistant cast iron handbook, American Foundry Society, Des Plaines, IL, 2000). On the graphitic side, grades such as Ni-Tensil, bainitic and martensitic cast irons with free graphite, Silal, Chugal, Duriron, Ni-Resist, Nicrosilal, Minvar and Nomag are compared in terms of strength, toughness, wear behavior, damping, thermal conductivity and corrosion resistance (ASM Handbook Committee, ASM handbook, volume 1: properties and selection: irons, steels, and high-performance alloys, ASM International, Materials Park, OH, 1990; Davis, ed., ASM specialty handbook: cast irons, ASM International, Materials Park, OH, 1996; Stefanescu, ed., ASM handbook, volume 1A: cast iron science and technology, ASM International, Materials Park, OH, 2017; Laird et al., Abrasion-resistant cast iron handbook, American Foundry Society, Des Plaines, IL, 2000). On the white-iron side, ledeburitic Ni-Hard, martensitic high-chromium grades and ferritic aluminum-alloyed irons without graphite are contrasted with respect to carbide type, hardness level, toughness and performance in severe abrasive, corrosive and high-temperature service (Davis, ed., ASM specialty handbook: cast irons, ASM International, Materials Park, OH, 1996; Stefanescu, ed., ASM

M. González Fernández de Castro et al., *Alloyed Cast Iron*, SpringerBriefs in Materials, https://doi.org/10.1007/978-3-032-27303-1_7

handbook, volume 1A: cast iron science and technology, ASM International, Materials Park, OH, 2017; Laird et al., Abrasion-resistant cast iron handbook, American Foundry Society, Des Plaines, IL, 2000; Fashu et al., Metals 13(11):1831, 2023). This overview is intended as a quick design and selection tool that links the detailed descriptions given in previous chapters with practical alloy choice for engineering components (ASM Handbook Committee, ASM handbook, volume 1: properties and selection: irons, steels, and high-performance alloys, ASM International, Materials Park, OH, 1990; Davis, ed., ASM specialty handbook: cast irons, ASM International, Materials Park, OH, 1996; Stefanescu, ed., ASM handbook, volume 1A: cast iron science and technology, ASM International, Materials Park, OH, 2017; Laird et al., Abrasion-resistant cast iron handbook, American Foundry Society, Des Plaines, IL, 2000; Fashu et al., Metals 13(11):1831, 2023).

Keywords Summary table · Classification · Alloyed cast irons · Free-graphite irons · White irons · Composition–microstructure–property relationships · Alloy selection

7.1 Summary Table

Quick reading note: free graphite = better damping and machinability; without graphite = white fracture, maximum hardness/wear or maximum oxidation depending on the carbide system/element (Cr or Al). The matrix (pearlitic/bainitic/martensitic/ferritic/austenitic) dictates the strength–toughness balance and the temperature behavior [1–5] (Fig. 7.1).

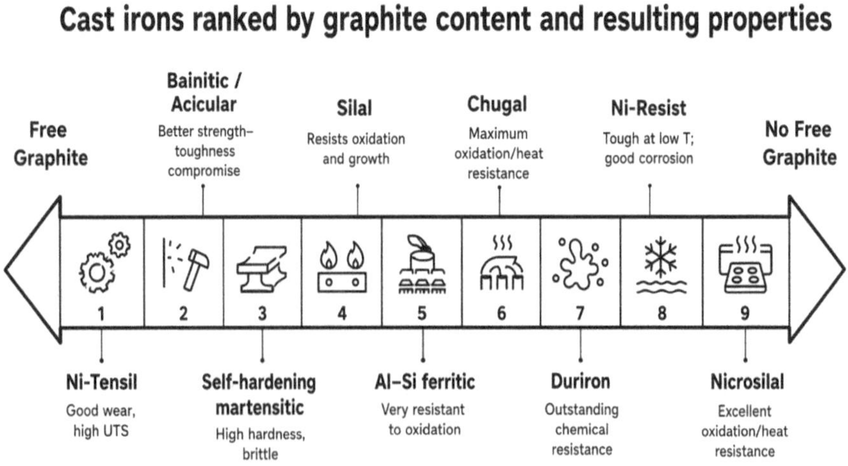

Fig. 7.1 Concise classification of alloyed cast irons according to their typical composition, room-temperature microstructure, notable properties, and main applications

Cast irons with free graphite

Type	Typical composition	Microstructure (RT)	Notable properties	Typical uses
Ni-Tensil (low-alloy pearlitic gray)	C 3.0–3.5; Si 1.5–2.9; Mn 0.6–0.9; Ni 0.5–3.5; Cr 0.3–1.0; Mo 0.2–1.0; Cu 0.2–1.0%	Flake graphite in fine pearlitic matrix	High UTS vs common grays; good wear; avoid free ferrite and massive carbides	High-strength parts where fine pearlite is desired (Ni-Tensil family)
Bainitic/acicular	Alloyed (e.g., Ni, Mo, Cu, Cr) for bainitic/acicular matrix	Free graphite with bainitic/acicular matrix	Better strength–toughness compromise than pearlitic; high UTS/toughness	Components seeking strength + toughness (acicular)
Self-hardening martensitic	Typical: C 2.9–3.3; Si 1.5–2.0; Mn 0.7–1.0; Cr 0.6–1.0; Ni 4.0–5.0%	Flake graphite + martensitic matrix with variable γR (self-hardens on mold cooling)	High hardness (≈400–500 HB after stated routes), brittle; temper/anneal options for machining	Wear without heavy blows; alternative to surface hardening when bulk hardening is desired
Silal (ferritic high Si)	Si 5.5–7.0%, C < 2.3, Mn 0.5–0.8 (opt. Cr ≤ 2, Mo ≤ 2)	Interdendritic graphite (Type D) in stable ferritic matrix; A1 ≈ 970 °C	Resists oxidation and growth up to 850–900 °C; 160–278 MPa, 140–190 HB; brittle cold, better >250 °C	Furnace grates/supports, nozzles, parts exposed to hot gases
Al–Si ferritic (4–6% Al)	C 2.5–3.5; Si 1.6–2.1; Al 4–6; Mn 0.1–0.4% (Mo up to 2% improves creep/oxidation)	Free graphite in ferritic matrix; possible nodular (bull's-eye) with Mg	Very resistant to oxidation (Al_2O_3); UTS 175–225 MPa (flake)/390–570 MPa (nodular); brittle; hard to machine	Hot furnace parts where oxidation rules and no severe shocks
Chugal (20–24% Al)	C ~ 2; Si ~ 1; Al 20–24%; sometimes Ce traces (modifier)	Free graphite in ferritic matrix; paramagnetic at >~20% Al	Maximum oxidation/heat resistance (1000–1100 °C); very brittle; Al_4C_3 risk if poorly processed	Extreme oxidation niches without mechanical impact; sulfurous/H_2SO_4 atmospheres favorable
Duriron (12–16% Si)	Si 12–16% (BS 14.25–15.25); C 0.4–1.0%	Interdendritic graphite in silico-ferrite; protective surface SiO_2 film	Outstanding chemical resistance (acids); very hard (450–520 HB) and brittle; UTS 90–180 MPa; requires stress-relief anneal; Durichlor variant (+~3% Mo) for HCl	Chemical drains, pumps/valves in chemical industry, cathodic-protection anodes

(continued)

Type	Typical composition	Microstructure (RT)	Notable properties	Typical uses
Ni-Resist (austenitic)	C 2–3; Si 1–6; Mn 0.5–2; Ni 13.5–36; Cr 1–6; Cu 0.5–7% (ductile: low Cu)	Austenite + graphite (flake or nodular); 5–10% carbides if high Cr	Tough at low T (no DBTT), good corrosion in seawater/alkalis; heat service up to 800–950 °C; weldable; high CTE except variants	Marine & chemical pumps/valving; engines (manifolds/turbines); cryogenics; precision (glass, etc.)
Nicrosilal (high-Si Ni-Resist)	Ni ~ 18–32; Si 4–6 (typ. 6); Cr 1.8–5; C 1.6–3%	Austenitic with Cr carbides for Cr > ~1.8; fine graphite preferable	Excellent oxidation/heat >700 °C; high CTE; more brittle than low-Cr Ni-Resist; beware 400–700 °C if (Si > 5, Cr < 1.8, Ni < 17) → hardens/embrittles	Furnace parts, melting molds, severe thermal service
Minvar (low CTE)	Ni 34–36; Si 1–2; C ≤ ~2.4; Mn 0.5–1.5; no Cu/Cr	Austenitic + graphite; very few carbides; ferromagnetic (Fe–36Ni)	Very low CTE ($\approx$3–6 × 10^{-6}/°C at 10–205 °C); soft (100–125 HB); good toughness even cryogenic	Glass molds, precision bases, dimensional compatibility with steel
Nomag (nonmagnetic, Ni + Mn)	C 2.5–3.0; Si 1.5–3.0; Mn 4–7; Ni 11–12 (ref. 10Ni–5Mn)	Austenitic + graphite; very low μr ($\approx$1.03–1.11); less tough/corrosion than high-Ni Ni-Resist	Nonmagnetic, high resistivity; mechanically weaker than high-Ni Ni-Resist; oxidation service $\approx\leq$500 °C	Parts in electric/magnetic fields; economical non-magnetic substitute

White cast irons (without free graphite)

Type	Typical composition	Microstructure (RT)	Notable properties	Typical uses
Ni-Hard (ledeburitic)	C 2.5–3.6; Si < 0.8 (up to 2.2 in Type 4); Ni 3.3–6.5; Cr 1.4–9; Mo < 1%	Eutectic carbides (M_3C or M_7C_3 in Type 4) in martensitic matrix (variable γR)	Very high hardness (500–780 DPN; Ni-Hard 4 > 700 HB HT); outstanding wear; brittle; temper 205–275 °C; cryogenic useful	Mills, balls, slurry pumps, crushers, rollers, augers, etc.
Hi-Cr 11–28% Cr (martensitic)	Cr 11–30; C 1.6–3.6; +Mo/Ni/Cu for hardenability	Discontinuous M_7C_3 + martensitic matrix (after HT; γR 10–30%)	450–550 HB as-cast → 600 HB–~67 HRC after HT; better toughness than Ni-Hard; good corrosion (matrix Cr > 12%) and heat 800–950 °C	Slurry pumping, mining, brick dies, pulverizers, etc.
>28% Cr (ferritic)	Cr 28–35; C 1.2–2.6; low + Mo/Ni/Cu	Ferritic (δ/α) matrix with Cr carbides ($M_7C_3/M_{23}C_6$); no graphite	450–600 HB as-cast (650–750 HB hardened; 350–400 HB annealed); very high oxidation/corrosion; brittle	Abrasive/corrosive slurries, furnace parts 800–900 °C
≈7% Al (ferritic Al, white)	Al ~ 7; C 2.5–3.0; Si ≤ 1; Mn ~ 0.5%	Ferrite + Fe_3AlC_x (κ phase) carbides; no graphite; very stable (no α ↔ γ)	Excellent oxidation (≈950 °C); 280–320 HB; UTS 100–180 MPa; brittle; density ~ 6.6 g/cm^3; machinable if C < 1.5%	Slurry pumps, furnaces, pulverizing/cement, drilling (abrasion + heat)

References

1. ASM Handbook Committee. (1990). *ASM handbook, volume 1: Properties and selection: Irons, steels, and high-performance alloys*. ASM International.
2. Davis, J. R. (Ed.). (1996). *ASM specialty handbook: Cast irons*. ASM International.
3. Stefanescu, D. M. (Ed.). (2017). *ASM handbook, volume 1A: Cast iron science and technology*. ASM International.
4. Laird, G., Gundlach, R., & Röhrig, K. (2000). *Abrasion-resistant cast iron handbook*. American Foundry Society.
5. Fashu, S., Xie, Y., Wang, J., & Xiao, W. (2023). Development and performance of high chromium white cast irons (HCWCIs) for wear/corrosive environments: A critical review. *Metals, 13*(11), 1831. https://doi.org/10.3390/met13111831

Chapter 8
State of the Art of Alloyed Ferrous Cast Irons Today

Abstract The alloyed ferrous castings sector is undergoing a phase of dynamic growth driven by increasing demand for high-quality steel and cast components, together with rapid advances in metallurgy, process control and digitalization (ASM International, ASM handbook, volume 1A: cast iron science and technology, ASM International, Materials Park, OH, 2017; Davis, ed., ASM specialty handbook: cast irons, ASM International, Materials Park, OH, 1996; World Foundry Organization, World Foundry Organization yearbook: global casting production and trends, 2020). This chapter reviews the current state of the art in alloyed ferrous cast irons and ferroalloys, from global market trends and key industrial applications to the implementation of Industry 4.0 concepts, automation and robotics, advanced simulation tools, metal additive manufacturing, and modern quality-control strategies (ASM International, ASM handbook, volume 1A: cast iron science and technology, ASM International, Materials Park, OH, 2017; Davis, ed., ASM specialty handbook: cast irons, ASM International, Materials Park, OH, 1996; World Foundry Organization, World Foundry Organization yearbook: global casting production and trends, 2020; Heine et al., Int J Metalcast 13(3):421–439, 2019). Particular emphasis is placed on how these technological developments interact with alloy design for wear-, corrosion- and heat-resistant cast irons, enabling higher performance, improved energy efficiency and reduced environmental impact (Davis, ed., ASM specialty handbook: cast irons, ASM International, Materials Park, OH, 1996; World Foundry Organization, World Foundry Organization yearbook: global casting production and trends, 2020; Heine et al., Int J Metalcast 13(3):421–439, 2019; Gibson et al., Additive manufacturing technologies: 3D printing, rapid prototyping, and direct digital manufacturing, 3rd ed., Springer, New York, 2021). Finally, emerging challenges and opportunities are discussed, including sustainability, circular-economy approaches, and the role of advanced alloyed irons in future manufacturing scenarios (World Foundry Organization, World Foundry Organization yearbook: global casting production and trends, 2020; Heine et al., Int J Metalcast 13(3):421–439, 2019; Gibson et al., Additive manufacturing technologies: 3D printing, rapid prototyping, and direct digital manufacturing, 3rd ed., Springer, New York, 2021).

M. González Fernández de Castro et al., *Alloyed Cast Iron*, SpringerBriefs in Materials, https://doi.org/10.1007/978-3-032-27303-1_8

Keywords Alloyed ferrous cast irons · State of the art · Global market · Industry 4.0 · Automation and robotics · Metal additive manufacturing · Quality control · Sustainability · Circular economy

8.1 Introduction

The sector of alloyed ferrous cast irons is in a period of significant transformation and growth, driven by the rising global demand for high-quality steel and ongoing advances in metallurgy. The ferroalloys market—essential components in the production of steel and alloyed metals—is experiencing notable expansion, with steady growth projections for gray cast iron and the overall iron casting market. For example, the global gray cast iron market, valued at USD 80,979.31 million in 2024, is expected to reach USD 102,561.75 million in 2033, with a compound annual growth rate (CAGR) of 2.66% from 2025 to 2033. More broadly, the iron casting market, valued at USD 151.770 million in 2024, is projected to grow to USD 269.700 million by 2033, with an impressive CAGR of 6.6% [1–5].

This growth is underpinned by the expansion of the steel industry, increasing global industrialization and urbanization, and the demand for high-quality steel in key sectors such as construction, automotive, energy, and heavy machinery. Technological advances in metallurgy, such as the use of electric arc furnaces, have improved the efficiency and cost-effectiveness of ferroalloy manufacturing, enabling the production of high-purity alloys for specialized applications [1–5].

A fundamental aspect of this evolution is the dual focus on high performance and sustainability. The industry not only seeks to improve material properties and process efficiency through advanced alloys and innovative manufacturing techniques, but also focuses on environmental responsibility. This is manifested in the adoption of greener production methods, the use of recycled materials, and the reduction of the carbon footprint. The need to balance superior performance with minimizing environmental impact drives integrated research and development, where materials innovation and ecological responsibility are interdependent [1–5].

Alloyed ferrous cast irons and ferroalloys form the backbone of numerous modern industries, acting as foundational materials that enable the manufacture of components with superior mechanical and chemical properties. Ferroalloys are alloys of iron with one or more additional elements, such as manganese, silicon, or chromium, which are crucial for producing high-quality steel and other alloyed metals. Their main function is to enhance characteristics such as strength, durability, corrosion resistance, and heat tolerance of base metals [1–5].

Ferrous casting parts—including cast iron and steel components—are widely used due to their superior strength, durability, and cost-effectiveness. The steady demand for industrial machinery, automotive parts, and infrastructure development ensures sustained growth in this market. The systemic importance of these materials is observed in their cross-sector application in high-demand areas such as construction, automotive, aerospace, and energy. Advances in this sector have a positive domino effect on global industrial development and technological progress. For example, any improvement in the properties or production efficiency of ferroalloys translates directly into more advanced components, more efficient manufacturing processes, and better performance across a wide range of products and infrastructures worldwide. This positions the ferrous casting industry not only as a materials supplier, but as a strategic enabler of overall industrial progress [1–5].

This report aims to provide a comprehensive overview of the current state of the art of alloyed ferrous cast irons. To this end, the following key aspects will be addressed: the global market landscape and its industrial applications, the most recent advances in materials and alloy metallurgy, innovations in manufacturing processes, advanced quality-control methodologies, sustainability initiatives and environmental impact, the main challenges facing the industry, and future perspectives in research and development. The structure of the report is designed to provide a comprehensive and systematic analysis, facilitating an understanding of the complex dynamics of this vital sector [1–5] (Fig. 8.1).

Alloyed Ferrous Cast Irons and Ferroalloys

Characteristic	Alloyed Ferrous Cast Irons	Ferroalloys
Market Growth	Significant transformation and growth	Notable expansion
Key Drivers	Rising global demand for high-quality steel, advances in metallurgy	Expansion of steel industry, industrialization, urbanization
Focus	High performance and sustainability	Enhancing strength, durability, corrosion resistance
Application	Foundational materials for components	Producing high-quality steel and alloyed metals
Importance	Superior strength, durability, cost-effectiveness	Positive domino effect on industrial development

Fig. 8.1 Comparative overview of alloyed ferrous cast irons and ferroalloys, contrasting market growth trends, key industrial drivers, strategic focus (performance and sustainability), main applications, and industrial importance

8.2 Global Market Landscape and Industrial Applications

The global market for alloyed ferrous cast irons is experiencing notable growth, driven by a combination of macroeconomic factors and technological advances. The ferroalloys market, which is critical in the production of steel and alloyed metals, is witnessing significant expansion. This trend is reflected in projections for specific industry segments. For example, the global gray cast iron market was valued at USD 80,979.31 million in 2024 and is projected to reach USD 102,561.75 million in 2033, with a CAGR of 2.66% from 2025 to 2033. More broadly, the iron casting market, valued at USD 151.770 million in 2024, is expected to grow consistently, reaching USD 161.780 million in 2025 and USD 269.700 million by 2033, with a robust CAGR of 6.6%. The overall iron and steel casting market generated USD 162.680 million in 2023 and is expected to register a CAGR of 3.8% from 2024 to 2032 [1–5].

The main drivers of this growth include the expansion of the steel industry—which accounts for the majority of ferroalloy consumption—and technological advances in metallurgy. Rising demand in emerging economies such as India, China, and Brazil, which are experiencing rapid industrial growth and increased investment in the construction and automotive sectors, also amplifies demand for ferroalloys [1–5].

The disparity in growth rates among different segments of ferrous casting—with a 6.6% CAGR for iron casting overall compared to 2.66% for gray cast iron or 3.8% for iron and steel casting in general—suggests robust and diversified demand. This observation indicates inherent resilience to localized economic fluctuations and a widespread industrial dependence on these materials. The higher growth rate for iron casting as a whole could be attributed to a strategic shift or greater preference toward more specialized, high-performance iron castings, such as ductile iron. The use of ductile iron, for example, has increased by 26% in heavy machinery and piping, which is identified as a key driver of the metal casting market. This trend underscores the importance of research and development in specialized iron alloys to meet the growing demands of industry [1–5].

Alloyed ferrous cast irons are indispensable materials across a wide range of industries, with demand driven by strategic sectors globally [1–5].

The automotive sector is one of the largest consumers of ferrous cast parts, using them in critical components such as engine blocks, brake drums, crankshafts, camshafts, and pistons. The growing demand for vehicles, especially electric vehicles (EVs), is transforming material requirements and thereby driving market growth. The production of high-performance cast-iron parts for battery housings and automotive enclosures has increased by 40% due to EV demand. This means not only a greater number of parts, but also a need for components with improved properties—such as lower weight, better thermal management, and higher corrosion resistance—which drives innovation in alloy design and casting processes [1–5].

In construction and infrastructure, iron and steel castings are fundamental for building bridges, buildings, pipes, and industrial machinery. Gray cast iron, for example, is widely used in pipes and fittings. The durability and corrosion resistance of ductile iron GGG50 make it ideal for underground and above-ground infrastructure projects, including pipes, fittings, and valves for water and wastewater systems [1–5].

The energy sector—particularly wind energy—relies heavily on alloyed ferrous castings. Components such as turbines, rotors, and shafts are manufactured from special ferrous alloys that require rigorous and specialized casting processes to meet performance and durability expectations in harsh environmental conditions, such as strong winds and temperature fluctuations. The ability of foundries to produce these components efficiently and reliably is vital for the continued growth of wind energy [1–5].

Heavy machinery and industrial equipment are also major consumers of ferrous castings. Components such as gears, gearboxes, and hydraulic parts often incorporate ductile iron GGG50 due to its excellent wear resistance and hardness, contributing to the longevity and reliability of equipment in demanding industrial environments [1–5].

Finally, in the mining industry, alloyed ferrous castings are crucial for components facing abrasive and challenging conditions, such as crusher parts, wear plates, and bucket teeth. Materials such as impact-resistant manganese steel and high-chromium alloy irons (Ni-Hard), with high resistance to abrasive wear, are examples of specialized castings used in this sector [1–5] (Fig. 8.2).

Alloyed Ferrous Cast Irons Market Overview

Characteristic	Gray Cast Iron	Iron Casting	Iron and Steel Casting
2024 Value (USD millions)	80,979.31	151,770	N/A
2033 Value (USD millions)	102,561.75	269,700	N/A
CAGR (2025-2033)	2.66%	6.6%	N/A
Key Drivers	Expansion of steel industry, technological advances, demand in emerging economies	Expansion of steel industry, technological advances, demand in emerging economies	Expansion of steel industry, technological advances, demand in emerging economies
Applications	Pipes and fittings	Engine blocks, brake drums, crankshafts, camshafts, pistons, pipes, fittings, valves, turbines, rotors, shafts, gears, gearboxes, hydraulic parts, crusher parts, wear plates, bucket teeth	Engine blocks, brake drums, crankshafts, camshafts, pistons, pipes, fittings, valves, turbines, rotors, shafts, gears, gearboxes, hydraulic parts, crusher parts, wear plates, bucket teeth

Fig. 8.2 Market overview table comparing gray cast iron, iron casting, and iron & steel casting sectors, summarizing 2024 and 2033 market values, CAGR (2025–2033), growth drivers, and representative industrial applications

8.3 Advances in Materials and Alloy Metallurgy

Advanced alloyed cast irons, particularly ductile iron, have achieved exceptional mechanical properties thanks to meticulous control of their microstructure. The combination of strength, ductility, and fatigue resistance in ductile iron is due to the morphology of the graphite (spheroidal) and the composition of the matrix phase [1–5].

There are different matrix phases that confer varied properties [1–5]:

- Ferritic ductile iron: With ferrite content equal to or greater than 90%, it offers high ductility (up to 20% elongation) and an ultimate tensile strength (UTS) of 350–450 MPa. It is ideal for components requiring high ductility, such as impact-absorbing housings [1–5].
- Pearlitic ductile iron: With pearlite content equal to or greater than 90%, it provides high strength (UTS of 650–800 MPa), although with limited elongation (6–8%). It is used in high-strength applications, such as gears and shafts [1–5].
- Mixed ferritic–pearlitic ductile iron: Offers a balance of phases (for example, 50:50) with a UTS of 400–550 MPa and an elongation of 10–15%, being suitable for general-purpose castings that combine strength and hardness [1–5].
- Austempered ductile iron (ADI): ADI variants can exceed 1000 MPa UTS, which makes them suitable for very high-strength applications [1–5].

Precise control of the microstructure is achieved through the addition of alloying elements and advanced heat treatments. For example, microalloying with copper and molybdenum (0.5–2.0% Mo) significantly improves the hardenability of high-chromium white irons, effectively suppressing the formation of pearlite and bainite. Research also focuses on reducing retained austenite and promoting the formation of martensite through sub-zero treatments, which leads to higher hardness. The ability to design specific microstructures through alloying and heat treatment allows manufacturers to optimize ferrous castings for highly specialized and demanding applications. This not only improves product reliability but also reduces material waste, taking foundry practice beyond simple shape making toward advanced materials engineering that seeks specific performance profiles through sophisticated metallurgical control [1–5].

The addition of specific alloying elements is fundamental to modify and optimize the properties of ferrous castings, enabling their use in increasingly demanding applications [1–5].

Niobium (Nb) is a crucial alloying element, with approximately 85–90% of its global production destined for the steel industry. Its incorporation can increase steel strength by more than 30% with additions of only 0.03–0.05%. Niobium influences solidification behavior by promoting grain refinement and reducing segregation, acting as nucleation sites and yielding uniform microstructures. However, excessive niobium can cause coarse precipitates or inclusions, which can generate casting defects such as porosity or hot tearing, requiring precise control of cooling rates and casting parameters to mitigate these issues. In addition to strength, niobium improves high-temperature oxidation resistance and corrosion resistance of steel, and optimizes its weldability [1–5].

Titanium (Ti), even in microalloying amounts, has a significant impact. It has been shown that microalloying with titanium effectively reduces residual stresses in synthetic gray cast iron, while maintaining tensile strength and Brinell hardness. It is also useful for mitigating the negative effects of nitrogen content, which can arise from the use of steel scrap in the production of synthetic cast iron. Titanium alloys are also employed in high-performance applications that demand a combination of strength and low weight, such as aerospace components [1–5].

Nickel (Ni) improves strength, ductility, impact resistance, and corrosion resistance in steel, even in acidic and alkaline environments. In ductile iron, nickel-magnesium master alloys are used to increase the magnesium content and enhance nodularity levels. Nickel also contributes to the formation of a pearlitic matrix and can promote graphitization in large amounts in gray iron [1–5].

Rare Earths (RE), often referred to as "industrial vitamins," possess irreplaceable magnetic, optical, and electrical properties. In metallurgy, they have been used for more than 30 years with mature technologies. The addition of rare-earth silicide or fluoride to iron and steel alloys can improve the processing performance of steel, as well as its desulfurization and refining. Ferrosilicon and silicon-magnesium alloys with rare earths are used as spheroidizing agents to produce nodular cast iron, especially suitable for complex parts in automotive, tractors, and diesel engines. The addition of rare earths can also improve the physical and chemical properties of other non-ferrous alloys [1–5].

High-Entropy Alloys (HEAs) represent a paradigm shift in alloy design, moving away from traditional methods that focus on a dominant base metal with minor additions. In contrast, HEAs incorporate five or more principal elements in nearly equiatomic proportions, generating high configurational entropy that stabilizes the crystal structure and results in superior mechanical strength and thermal stability [1–5].

These alloys exhibit unique and enhanced properties, such as high strength, hardness, ductility, corrosion resistance, wear resistance, thermal stability, and high-temperature creep resistance. They can also present improved thermal and electrical conductivity, as well as specific magnetic properties. The ability to operate at higher temperatures than conventional alloys makes them candidates to replace traditional nickel-based superalloys in high-temperature applications [1–5].

The fundamental mechanisms conferring these outstanding properties are the high-entropy effect (which promotes formation of simple solid solutions), severe lattice distortion (due to atomic size differences and bonding energies), sluggish diffusion (which hinders atomic motion and improves high-temperature stability), and the cocktail effect (the synergistic combination of multiple elements) [1–5].

Although HEAs are already applied in high-tech industries such as aerospace, defense, nuclear, and biomedical, their specific application in ferrous castings is still at an early research stage. However, the high-entropy concept is being applied to modify high-chromium white irons, creating "high-entropy high-chromium hybrid cast irons" that show superior sliding wear resistance due to their high hardness.

Comparison of Ductile Iron Matrix Phases

Property	Ferritic Ductile Iron	Pearlitic Ductile Iron	Mixed Ferritic-Pearlitic Ductile Iron	Austempered Ductile Iron (ADI)
Ferrite Content	≥90%	~10%	Variable	Not applicable
Pearlite Content	~10%	≥90%	Variable	Not applicable
UTS (MPa)	350–450	650–800	400–550	1000
Elongation (%)	Up to 20	6–8	10–15	Variable
Ideal Use	Impact-absorbing housings	Gears and shafts	General-purpose castings	Very high-strength applications

Fig. 8.3 Comparison of ductile iron matrix phases (ferritic, pearlitic, mixed ferritic–pearlitic, and austempered ductile iron—ADI), relating ferrite/pearlite content with UTS, elongation, and typical engineering uses

Research in this field is promising, and HEAs are expected to open new possibilities in the design of advanced materials for ferrous castings, offering benefits in terms of cost, energy, and sustainability. Current challenges include microstructural phase segregation and solidification cracking due to the varying melting points of the constituent elements, which often requires rapid solidification to avoid physical segregation. Modeling and AI-based tools are essential to efficiently explore and optimize these new alloys [1–5] (Fig. 8.3).

8.4 Innovations in Manufacturing Processes

The ferrous foundry industry is continuously adopting and developing new molding and casting techniques to improve the precision, efficiency, and complexity of the parts produced [1–5].

Among the most prominent molding techniques are [1–5]:

- Sand molding: Remains the most common method due to the abundance and low cost of sand. It is ideal for producing large, robust metal parts. Innovations include the use of finer sand mixed with heated resin to form a hardened shell (shell molding), which enables a finer surface finish and is ideal for more complex parts [1–5].
- Permanent mold (metal mold) casting: Uses molds made of metal with a melting point higher than that of the metal to be cast. Although it cannot produce very

fine shapes, it allows the size and shape of the final casting to be predicted with high precision, resulting in dense, fine-grained parts with a smoother surface finish [1–5].

- Continuous casting: Refines the casting process for serial production. The mold is open and water-cooled, and the casting passes through different processes before completion, being more economical and efficient for manufacturing standard products [1–5].
- Centrifugal casting: Involves introducing liquid metal into a rotating mold cooled by water. The centrifugal force creates a uniform layer of metal, although it only works with uniform-geometry shapes and is used for small parts, such as jewelry [1–5].
- Die casting: Consists of forcing molten metal into mold cavities at high pressure. Although commonly used with non-ferrous metals such as zinc, copper, and aluminum alloys, it produces well-detailed surfaces [1–5].
- Investment casting (lost-wax): Also known as lost-wax casting, it is a process that begins with a wax pattern of the desired part, which is then coated with a clay mold. The wax pattern melts during firing, leaving only the ceramic mold. This method is ideal for complex, high-precision parts, commonly used in power-generation, automotive, and aerospace components [1–5].

Innovations in these techniques, such as mobile foundry systems that use interconnected units to pour molten alloys on demand, optimize supply routes and enable more flexible, efficient, and sustainable production. The use of mobile crucibles reduces the need for manual pouring, minimizing the formation of oxides and slag, which results in cleaner and stronger final products [1–5].

Automation and robotics are revolutionizing the ferrous foundry industry, enabling more flexible, efficient, and safe production. Adoption of automation in foundries has reached 61% in North America and 41% in Europe, with increasing integration of robotics and CAD simulations [1–5].

Key applications of automation and robotics include [1–5]:

- End-to-end process automation: Solutions covering the entire production line, from the fast trimming press to cooling stacks, within a fully customized cell concept [1–5].
- Robots for foundry: Robots designed with heat-resistant special steel, capable of handling loads from 5 to 1300 kg, ideal for high-temperature environments in foundry production [1–5].
- Efficiency in high-temperature environments: New robot models offer high efficiency even under the complex conditions of the foundry and forging industry [1–5].
- Smart production with data exchange: Integration of OPC UA servers in robots and casting machines enables simple, platform-independent, scalable, internet-compatible data exchange, ensuring data availability across all systems at all times [1–5].
- Effective control through Big Data analytics: Cloud-based platforms for digital data acquisition in production allow access to robot data in real time, facilitating

Big Data analysis to control productivity and predict needed maintenance tasks [1–5].
- Casting cells 4.0: Precise interaction between robots, presses, and casting turntables, where cycle times are automatically optimized based on production pace, reducing waiting times and machine wear. Robot and press parameters are integrated into a single control unit for a global system view [1–5].

The benefits of these automated solutions are multiple: greater transparency and overall data visibility, increased production flexibility, high precision in complex processes (including handling demanding materials such as aluminum castings), and permanently reproducible high quality. Specific examples include automating die casting for automotive components, fast and efficient deburring of metal parts, and gravity casting [1–5].

Metal additive manufacturing, commonly known as metal 3D printing, represents a disruptive innovation in the production of metal parts. This technology enables creating three-dimensional objects directly from 3D CAD data, building structures by fusing thin successive layers of metal powder [1–5].

The main advantage of additive manufacturing is its ability to produce complex shapes that were impossible to achieve with traditional casting and machining methods (subtractive techniques). This includes intricate geometries, internal channels for conformal cooling, hidden features, thin walls, and lattices. The technology offers unprecedented design freedom, overcoming the constraints of conventional methods [1–5].

Additive manufacturing systems operate with a variety of metals, including titanium alloys (Ti6Al4V), cobalt-chrome, stainless steel, nickel alloys (Inconel 625 and Inconel 718), and aluminum alloys (AlSi10Mg). The process involves depositing a uniform layer of metal powder onto a build plate, followed by selective melting of areas with a high-power ytterbium fiber laser. This process is repeated layer by layer until the part is complete [1–5].

Additional benefits of additive manufacturing include [1–5]:

- Part consolidation: Allows reducing the number of elements in an assembly to a single complex component [1–5].
- Lower tooling cost: Parts are made directly, eliminating the need for expensive tooling [1–5].
- Weight reduction: Material is used only where necessary, allowing significant weight optimization [1–5].
- Customization: Facilitates the production of bespoke, made-to-order parts [1–5].
- Rapid design iterations: Enables rapid prototyping and direct-to-manufacturing [1–5].
- Complementary tool: Can be integrated into existing manufacturing processes to reduce steps, time-to-market, and costs [1–5].

Early adopters of this technology have been high-end sectors such as automotive, aerospace, and consumer goods, and its applications are rapidly expanding to the dental, medical, and tooling sectors. Additive manufacturing is expected to be

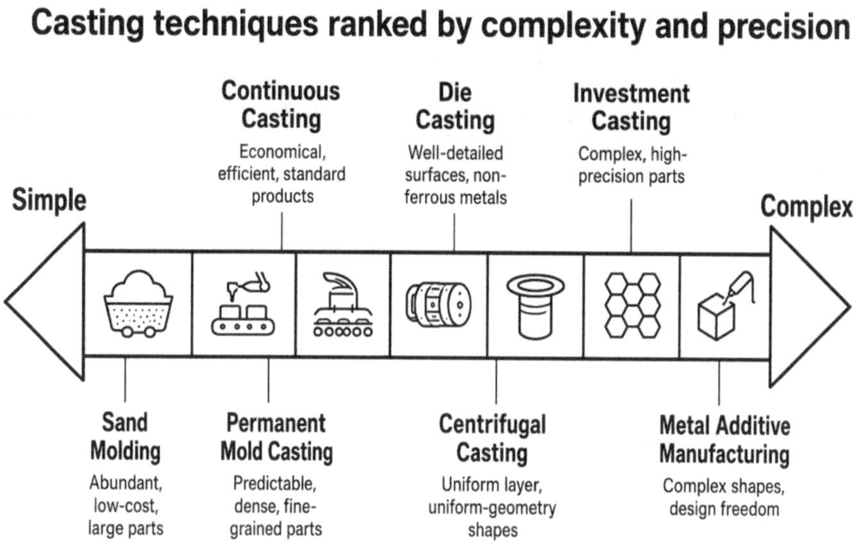

Fig. 8.4 Ranking of casting techniques by complexity and precision, illustrating the progression from sand molding and permanent mold casting to die casting, investment casting, and metal additive manufacturing in terms of achievable geometry and surface quality

integrated across all manufacturing sectors, marking a milestone in the evolution of metal production [1–5] (Fig. 8.4).

8.5 Advanced Quality Control and Sustainability

Quality control in ferrous foundries is crucial to ensure the integrity, performance, and reliability of parts. Advanced methods range from rigorous inspection and testing to statistical process control and defect prevention [1–5]

- Inspection and Testing: Inspections are carried out both in-process and at final stage to detect defects early and ensure the casting meets all specifications. Advanced inspection systems include [1–5]:
  - Vision systems and Coordinate Measuring Machines (CMMs): These technologies reduce labor costs and processing time while improving the accuracy and speed of measurements. CMMs operate via programmed routines in which engineers send commands to a control unit that moves a probed tool to gather precise measurements of the part [1–5].
- Non-Destructive Testing (NDT): Cost-effective methods that preserve part integrity by inspecting internal quality without damage. They include [1–5]:

  - Radiography: Uses X-rays (e.g., 100–400 kV) to detect internal defects such as cracks or porosity by passing a beam through the part to create an image. Common for aluminum castings [1–5].
  - Ultrasonic testing: Uses sound waves (e.g., 2–10 MHz) to identify defects on the internal surface of metals [1–5].
  - Eddy current testing: Identifies surface and near-surface defects in conductive materials [1–5].

- Destructive Testing: Used on samples to verify the strength of cast parts by applying high loads to fracture. This determines performance under intense stress and includes [1–5]:
  - Tensile tests: Measure the force a component can withstand before breaking (e.g., 100–1000 MPa) [1–5].
  - Impact tests: Record a part's ability to withstand and resist sudden blows (e.g., 10–100 J) [1–5].
  - Hardness tests: Measure a metal's hardness or softness (e.g., 50–100 HRB) [1–5].

8.5.1 Process Control and Defect Prevention

Constant monitoring of key process parameters is essential to produce uniform parts. This includes controlling melting points (around 680–720 °C for die casting), injection pressure (100–150 MPa), cooling temperature (up to 10–20 °C/s), and cycle time (30–60 s) [1–5].

Statistical Process Control (SPC) is integrated to monitor and improve processes using control charts and process capability analysis. In die casting, SPC can control injection pressure, cooling rates, and melt temperatures, helping prevent variation in the final product [1–5].

Proactive defect-prevention measures include [1–5]:

- Regular maintenance: Keep equipment and tooling in good condition to avoid defects and downtime [1–5].
- Quality management systems: Apply systems such as ISO 9001 to ensure consistent quality [1–5].
- Temperature control: Use advanced furnaces that precisely regulate temperature to avoid incomplete melting (too low) or excessive oxidation (too high) [1–5].
- Mold design and quality: Mold design and condition are essential; carry out regular inspections for wear and replace damaged molds immediately [1–5].
- Pouring techniques: Pay attention to molten metal flow, using techniques such as gating and risering to ensure the metal fills the mold evenly [1–5].
- Simulation and modeling: Simulation software (e.g., PoligonSoft) enables a digital twin of the mold and simulates each critical stage of the casting process,

identifying and preventing defects such as shrinkage cavities, residual stresses, hot tears, and distortion before physical production. This reduces equipment retuning costs and time, minimizing trial pours [1–5].

Common casting defects include surface defects (rough surfaces, "elephant skin"), inclusion defects (slag, sand), molding and pouring defects (misruns, cold shuts, sand instability), and cooling defects (shrinkage porosity, hot tearing, warpage). Applying these quality-control measures can reduce scrap rates by up to 10–20% and improve overall yield by 5–10%, translating into lower production costs and consistent part quality [1–5].

The ferrous foundry industry is under increasing regulatory pressure and greater consumer awareness to adopt more sustainable practices and reduce environmental impact. This has led to a strong emphasis on greener production methods and energy-use optimization [1–5].

8.5.2 *Environmental Impacts*

Iron and steel manufacturing, including foundries, generates various environmental impacts [1–5]:

- Air emissions: Large amounts of emissions occur during thermal, physical, and mechanical processes. These include dust and particulates (from melting furnaces, raw-material handling, shakeout), nitrogen oxides (NOx) and sulfur oxides (SOx), carbon monoxide (CO), volatile organic compounds (VOCs), and in some cases dioxins and furans. Cupola furnaces and electric-arc furnaces are significant sources of particulates and CO [1–5].
- Greenhouse-gas (GHG) generation: Casting is energy-intensive, and CO_2 emissions from fuel combustion are the main contributors to GHGs. Melting accounts for 40–60% of total energy consumption [1–5].
- Solid waste: Foundries generate large volumes of waste, including spent sands (65–80% of the total in ferrous foundries), desulfurization and melting slags, dust collected from emission-control systems, and refractory waste. Some may contain heavy metals (zinc, lead, cadmium, chromium, nickel) and be classified as hazardous [1–5].
- Wastewater: Large volumes come from cooling systems, core making, high-pressure die casting, and finishing operations. These waters may contain heavy metals, organics, oils, and greases [1–5].

8.5.3 Prevention and Control Measures

The industry is implementing multiple strategies to mitigate these impacts [1–5]:

- Reducing air emissions [1–5]:
  - Use pneumatic conveying and dust-controlled conveyors to reduce fugitive emissions [1–5].
  - Install extraction hoods and high-efficiency dust-collection systems (bag filters, cyclone collectors) on furnaces and process areas [1–5].
  - Employ low-NOx burners and optimize air/fuel ratios [1–5].
  - Use low-sulfur fuels, such as natural gas, and wet scrubbing systems [1–5].
  - Post-combustion of exhaust gases at high temperatures (>1200 °C) followed by rapid quenching to minimize dioxin formation [1–5].

- Optimizing energy consumption [1–5]:
  - Replace conventional cupolas with induction furnaces or cokeless cupolas, which are cleaner and more efficient [1–5].
  - Implement heat-recovery systems from industrial processes, such as waste heat from air compressors to heat water and furnaces, which can avoid significant CO_2 emissions [1–5].
  - Use variable-frequency drives on motors to reduce electricity use, with savings up to 50% on peak consumption and in fluid regulation [1–5].
  - Preheat scrap before use [1–5].
  - Closely monitor energy consumption to identify improvement areas [1–5].
  - Solid-waste management and circular economy [1–5]:

    Maximize on-site sand reuse and consider external reuse (e.g., in concrete or construction fill) [1–5].

    Recirculate baghouse dust back to furnaces to recover metals and minimize landfill waste [1–5].

    Minimize slag generation by process optimization and reuse slag (e.g., in block making or road base) [1–5].

    Raw-material shortages, exacerbated by global events such as the COVID-19 pandemic and geopolitical conflicts, have driven adoption of circular-economy models. Reuse and recycling of materials, such as high-quality scrap (which makes up 40% of recycled materials in foundries), are key solutions to cut costs and dependence on virgin inputs. Over 45% of foundries already use recycled cast-iron materials, reducing carbon emissions by 35% [1–5].
- Wastewater treatment [1–5]:
  - Install closed-loop circuits for cooling water to reduce consumption and discharge [1–5].
  - Recycle shakeout water by settling or centrifuging and filtering [1–5].
  - Treat industrial wastewater through chemical precipitation, coagulation, flocculation, and filtration to remove heavy metals and suspended solids [1–5].
  - Pave process areas and segregate contaminated stormwater for treatment [1–5].

Quality Control and Sustainability in Ferrous Foundries

Characteristic	Quality Control	Sustainability
Focus	Integrity, performance, reliability	Greener production, energy optimization
Methods	Inspection, testing, SPC	Emission reduction, energy optimization
Technologies	Vision systems, CMMs, NDT	Induction furnaces, heat recovery
Benefits	Lower costs, consistent quality	Regulatory compliance, brand image

Fig. 8.5 Integrated diagram of quality control and sustainability strategies in ferrous foundries, summarizing focus areas (integrity and greener production), methods (inspection, testing, SPC), technologies (vision systems, CMM, NDT, induction furnaces, heat recovery), and resulting benefits (cost reduction, consistency, regulatory compliance)

The emphasis on green production methods is emerging as a significant growth driver. Increasing regulatory pressure and consumer awareness of environmental sustainability are forcing manufacturers to adopt more environmentally friendly processes. Companies that invest in sustainable practices not only comply with regulations but also attract environmentally conscious customers and stakeholders, thereby improving brand image and market competitiveness [1–5] (Fig. 8.5).

8.6 Challenges and Future Outlook

Despite growth and technological advances, the alloyed ferrous foundry industry faces several significant challenges that require strategic attention [1–5].

- Raw-material price volatility: Raw-material costs account for a substantial portion of production expenses (up to 50% in the cast-iron industry). Price fluctuations—such as a 30% increase in iron ore costs over the past 2 years due to supply-chain disruptions and geopolitical instability—directly impact profit margins. The shortage of high-quality scrap, crucial for recycling and sustainability, also strains the supply chain. This volatility forces manufacturers to seek alternative materials, which may affect component durability and performance [1–5].

- Environmental compliance and regulations: Strict environmental regulations and growing consumer preference for sustainable manufacturing drive the need for greener foundry processes. However, compliance can hinder small foundries. Foundries must invest in emission-reduction technologies and waste-management systems to meet standards, which can entail significant costs [1–5].
- Challenges in cast-iron welding: Cast iron presents unique welding challenges due to its high carbon content (2–4%), which can lead to brittle structures such as martensite and increase the risk of cracking. Graphite flakes in gray iron act as stress concentrators, and the inherently low ductility of cast iron makes it difficult to accommodate expansion-contraction stresses during welding. Impurities such as sulfur and phosphorus can also degrade weldability [1–5].
- Casting defects: Despite advances in quality control, casting defects remain a challenge. These include surface defects (roughness), inclusion defects (slag, sand), molding and pouring defects (misruns, cold shuts, sand instability), and cooling defects (shrinkage porosity, hot tearing, distortion). Simulation and modeling help predict them, but complete elimination remains an ongoing goal [1–5].
- The future of alloyed ferrous castings looks promising, driven by continuous innovation and a strong focus on sustainability and digitalization [1–5].
- Lightweight materials and high-performance alloys: The demand for lighter vehicles to improve fuel efficiency and reduce emissions is driving a shift toward lightweight materials. Although aluminum and magnesium alloys are gaining ground, ferrous castings will remain essential. Continued development of high-performance alloys is expected to meet the demanding requirements of modern vehicles and other applications, improving overall performance and reliability. This includes further research into high-entropy alloys to surpass conventional material properties [1–5].
- Advanced simulation and modeling technologies: Simulation and modeling will remain critical tools to optimize mold design, predict and prevent defects, reduce material consumption, and improve final product quality. The accuracy of these tools reduces waste, saves time, and lowers production costs [1–5].
- Sustainability and circular economy: Sustainability will be a central focus, with the industry adopting greener practices through the use of recycled materials and the development of more energy-efficient processes. Integrating the circular economy—maximizing material reuse and recycling—will be fundamental to reducing dependence on virgin raw materials and mitigating environmental impacts [1–5].
- Automation, robotics, and Industry 4.0: Automation and robotics will continue to transform the industry, with robotic systems increasingly used for tasks such as pouring molten metal, removing castings from molds, and quality inspections. The integration of Industry 4.0 technologies, such as the Internet of Things (IoT) and artificial intelligence (AI), will further drive efficiency and innovation in

metal casting. Digital foundries—already 29% of the total operating base—will significantly improve result accuracy and reduce defects [1–5].

- Innovations in melting and post-treatment processes: Advances are expected in cleaner, more efficient melting techniques, such as induction furnaces. Spectroscopy, for example, is already used to optimize casting processes by enabling precise control of molten-metal composition, minimizing the need for reprocessing and reducing costs. Advanced heat treatments, including sub-zero treatments, will remain key to fine-tuning the mechanical and microstructural properties of alloys [1–5].
- Mobile casting and on-demand production: The emergence of mobile casting, with interconnected units that pour molten alloys on demand, will optimize supply routes and enable more flexible and efficient production [1–5].

These developments point to a smarter, more efficient, sustainable alloyed ferrous foundry industry, tailored to the demands of high-performance materials in an ever-evolving industrial landscape [1–5].

8.7 Conclusions

The alloyed ferrous foundry sector is in a dynamic phase of evolution, driven by the growing global demand for high-performance materials and the pressing need for sustainability. The state-of-the-art analysis reveals an industry not only experiencing steady growth across multiple segments, but also proactively adapting to market challenges and environmental requirements [1–5].

Demand for high-quality steel—particularly in sectors such as automotive (including electric vehicles), construction, energy, and heavy machinery—acts as a primary engine for growth in the ferrous casting market. This demand spurs metallurgical innovation, leading to the creation of alloys with superior mechanical and microstructural properties, such as ductile iron optimized for high-strength and fatigue applications [1–5].

Technological advances are evident at every stage of manufacturing. New molding techniques, automation and robotics (integrated with Industry 4.0, IoT, and AI), and metal additive manufacturing are transforming production, enabling more complex, precise, and efficient parts. In parallel, quality-control methods have become more sophisticated, employing non-destructive inspection tools and statistical process control to ensure component integrity and performance [1–5].

Sustainability has become a strategic pillar, with a strong focus on reducing emissions, optimizing energy use, and implementing circular-economy models. The reuse of materials and scrap recycling not only address raw-material shortages but also contribute to a smaller carbon footprint [1–5].

Although challenges persist—such as raw-material price volatility and the complexity of welding certain alloys—the industry's future direction is clear: increased investment in R&D for advanced materials (including high-entropy alloys),

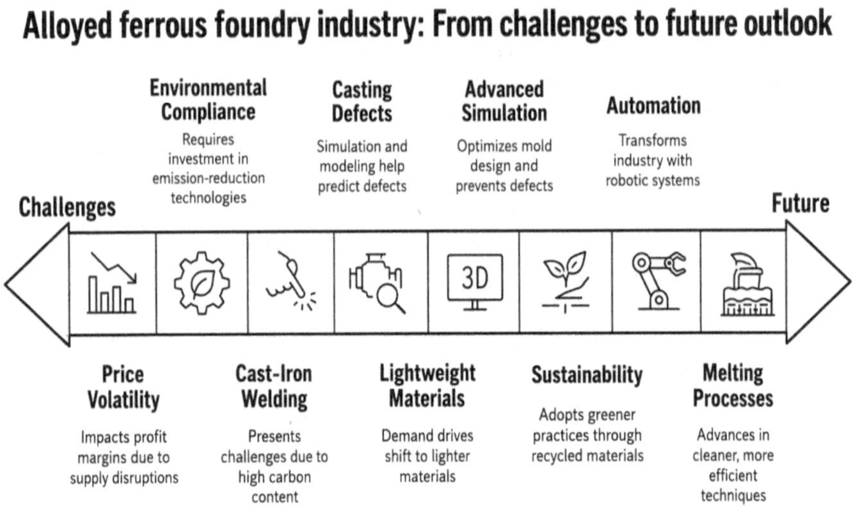

Fig. 8.6 Strategic outlook diagram for the alloyed ferrous foundry industry, linking current challenges (environmental compliance, casting defects, advanced simulation needs) with future directions such as volatility management, lightweight alloys, sustainability, automation, and digital foundry development

comprehensive process digitalization, and a continued search for more sustainable and efficient solutions. Alloyed ferrous foundries are not only meeting current needs, they are also laying the groundwork for the next generation of materials and industrial processes [1–5] (Fig. 8.6).

References

1. ASM International. (2017). *ASM handbook, volume 1A: Cast iron science and technology*. ASM International.
2. Davis, J. R. (Ed.). (1996). *ASM specialty handbook: Cast irons*. ASM International.
3. World Foundry Organization. (2020). *World Foundry Organization yearbook: Global casting production and trends*.
4. Heine, B., et al. (2019). Foundry 4.0: Digitalisation, automation and data-driven optimization in modern foundries. *International Journal of Metalcasting, 13*(3), 421–439.
5. Gibson, I., Rosen, D., & Stucker, B. (2021). *Additive manufacturing technologies: 3D Printing, rapid prototyping, and direct digital manufacturing* (3rd ed.). Springer.

FSC
www.fsc.org
MIX
Papier | Fördert
gute Waldnutzung
FSC® C083411

Zeitfracht Medien GmbH
Ferdinand-Jühlke-Straße 7
99095 Erfurt, Deutschland
produktsicherheit@kolibri360.de